CAMBRIDGE STUDIES IN LOW TEMPERATURE PHYSICS

EDITORS

Professor A.M. Goldman
Tate Laboratory of Physics, University of Minnesota

Dr P.V.E. McClintock
Department of Physics, University of Lancaster

Professor M. Springford
Department of Physics, University of Bristol

An introduction to millikelvin technology

An introduction to millikelvin technology

DAVID S. BETTS

Reader in Experimental Physics at the University of Sussex

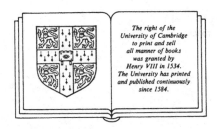

The right of the
University of Cambridge
to print and sell
all manner of books
was granted by
Henry VIII in 1534.
The University has printed
and published continuously
since 1584.

CAMBRIDGE UNIVERSITY PRESS

CAMBRIDGE

NEW YORK   NEW ROCHELLE   MELBOURNE   SYDNEY

CAMBRIDGE UNIVERSITY PRESS
Cambridge, New York, Melbourne, Madrid, Cape Town, Singapore, São Paulo

Cambridge University Press
The Edinburgh Building, Cambridge CB2 2RU, UK

Published in the United States of America by Cambridge University Press, New York

www.cambridge.org
Information on this title: www.cambridge.org/9780521344562

© Cambridge University Press 1989

This book is in copyright. Subject to statutory exception
and to the provisions of relevant collective licensing agreements,
no reproduction of any part may take place without
the written permission of Cambridge University Press.

First published 1989
This digitally printed first paperback version 2005

A catalogue record for this publication is available from the British Library

Library of Congress Cataloguing in Publication data

Betts, D. S. (David Sheridan), 1936–
An introduction to millikelvin technology.
(Cambridge studies in low temperature physics)
Bibliography: p.
Includes index.
1. Low temperatures.
2. Thermometers and thermometry.
I. Title.
II. Series.
QC278.B46 1989   536'.56   87-30928

ISBN-13 978-0-521-34456-2 hardback
ISBN-10 0-521-34456-5 hardback

ISBN-13 978-0-521-01817-3 paperback
ISBN-10 0-521-01817-X paperback

Contents

Preface   vii

1 Introduction to refrigeration and thermometry   1
  1.1 Preamble   1
  1.2 Refrigeration   1
    1.2.1 Free expansion of a fluid   1
    1.2.2 Isentropic expansion or compression of a fluid 2
    1.2.3 Isenthalpic expansion of a fluid   5
    1.2.4 Adiabatic (isentropic) demagnetisation of a
        paramagnet   7
  1.3 Thermometry   10
    1.3.1 The Kelvin scale   10
    1.3.2 International practical temperature scales   12

2 Properties of fluid ^{3}He/^{4}He mixtures   14
  2.1 Preamble   14
  2.2 Phase diagrams   14
  2.3 Dilute mixtures   17
  2.4 Fermi degeneracy of solute helium-3   17
  2.5 Mixtures and the two-fluid model   18
  2.6 Osmotic pressure   18
  2.7 Vapour pressure   21
  2.8 Transport properties   21

3 Dilution refrigeration   24
  3.1 Preamble   24
  3.2 Evaporation cooling   24
  3.3 Layout of components in a dilution refrigerator   25
  3.4 Startup   30
  3.5 Amount and concentration of mixture   31
  3.6 The still   31

3.7 How to obtain the lowest temperatures   34
3.8 Heat exchangers   34
3.9 Heat leaks   41
3.10 Construction of heat exchangers   43
3.11 Alternative methods avoiding the need for
     exchangers   43

4 The Pomeranchuk refrigerator   46
4.1 Preamble   46
4.2 Properties of melting helium-3   48
4.3 Cooling by solidification   49
4.4 Need to be gentle in compression   52
4.5 Some practical designs   54
4.6 Some more recent designs   57
4.7 Conclusions   59

5 Adiabatic nuclear demagnetisation   60
5.1 Preamble   60
5.2 Basic ideas   61
5.3 Entropy data   62
5.4 Ideal nuclear paramagnet   64
5.5 Reality (non-ideality)   65
5.6 Spin–lattice relaxation   66
5.7 Hyperfine-enhanced Van Vleck paramagnets   69
5.8 Geometry of the refrigerant: plates, wires, or powder?   70
5.9 Apparatuses   73

6 Thermometry   81
6.1 Preamble   81
6.2 The NBS superconducting fixed point device   83
6.3 Vapour pressure of helium-3   84
6.4 Melting pressure of helium-3   85
6.5 Carbon or germanium resistance   87
6.6 Capacitance   89
6.7 Cerous magnesium nitrate (CMN and CLMN)   89
6.8 NMR methods   91
6.9 Gamma-ray anisotropy   94
6.10 Noise thermometry   95
6.11 Conclusion   95

References   96
Index   101

Preface

The origin of this volume was an invitation I received from Dr Marek Finger of the Charles University, Prague, and the Joint Institute for Nuclear Research in Dubna, Russia, to give four lectures on low temperature methods at an international summer school on hyperfine interactions and physics with oriented nuclei organised at a chateau in Bechyně in the Czechoslovakian countryside, in September 1985. The topic of the summer school was something I knew little about, but low temperature physics is my métier and the preparation of the lectures was frankly not a large task, particularly in view of the fact that I was already the author of *Refrigeration and thermometry below one kelvin* (Sussex University Press, 1976). I decided to use a minimum of prose, produced in the usual garish colours, together with a large number of diagrams from various sources converted into transparencies. I would depend on my knowledge of the subject matter to talk through the transparencies in an unscripted way. It took me four days to think through the content and prepare the material. All the lectures were given on 3 September 1985. The organising committee originally had no intention of publishing proceedings but many participants expressed their desire to have the lectures and contributions presented in written form. My heart sank at the thought of converting my bundle of transparencies into something which could fairly be described as a camera-ready manuscript, but I agreed to try. It was like trying to turn a movie into a novella. I worked spasmodically on it, missing all of a series of extended deadlines until eventually the editors gave up on me and the proceedings appeared without my contribution. By this time I felt that I might as well finish what I had begun, and the *Four lectures on low temperature methods* appeared as a sort of private edition, available to some Sussex post-graduates but otherwise unknown. At this stage (August 1986) I sent a copy to Dr Simon Capelin of CUP with whom I had earlier been in correspondence about another project and offered to undertake a further but limited expansion of the work into the present book form. I had half expected this proposal to be rejected, but in fact some very helpful suggestions were made by Dr Peter McClintock and I was encouraged to proceed. By this time I had all the written material on word-processor disks and the rearrangements and expansion progressed steadily until I submitted the manuscript in June 1987. The integrated amount of work over the whole period was far more than I would ever have agreed to as a package, but because it had three distinct stages it always seemed a manageable task.

I mention all this because the final form has been to a large extent determined by its history. It is short, and it has a high figure/text ratio. The figures are not intended merely as adjuncts to the text; rather it is intended that the reader should spend time absorbing the significance of each before moving on. Also it is important to realise that it is in no way intended as a substitute for my *Refrigeration and thermometry below one kelvin* (1976) or for O.V. Lounasmaa's *Experimental principles and methods below 1 K* (1974), both of which are still available as more advanced sources.

This short work is intended as an introduction to the experimental technicalities of low and ultralow temperature physics research. It is likely to be of greatest value firstly to those who are beginning such research either as postgraduate students or as seasoned researchers moving in from another field, and secondly to final-year undergraduates choosing a low temperature physics option. There is a deliberate attempt to use diagrams as aids to understanding, and to refer readers to the professional literature as soon as the level of understanding is high enough.

DAVID S. BETTS

1 Introduction to refrigeration and thermometry

1.1 Preamble

It is useful when learning about the technicalities of low temperature physics to have ready access to some fundamental thermodynamic notions relating to refrigeration and thermometry. These can readily be summarised and that is the purpose of this chapter.

1.2 Refrigeration

The physical systems commonly exploited as refrigerants in low temperature research are fluids (specifically ^{3}He, ^{4}He, and ^{3}He–^{4}He mixtures), solids with electronic paramagnetism (for example, cerous magnesium nitrate (CMN)) or nuclear paramagnetism (for example, copper) or hyperfine-enhanced nuclear paramagnetism (for example, praesodymium nickel five (PrNi$_5$)), and solid/liquid ^{3}He at its melting pressure. For present illustrative purposes, however, we shall consider the thermodynamics of refrigeration using a monatomic van der Waals fluid; and using an idealised paramagnet.

1.2.1 Free expansion of a fluid

The fluid expands into a previously evacuated space and if the container has walls which are perfectly insulating and perfectly rigid then it is trivial to show that the internal energy U remains constant. The process is irreversible (entropy rises), and in the single-phase state the temperature change can conveniently be expressed in the following form:

$$\left(\frac{\partial T}{\partial V}\right)_U = -\frac{1}{C_V}\left[T\left(\frac{\partial p}{\partial T}\right)_V - p\right]. \qquad (1.1)$$

This is in general not zero (except for a perfect gas) but can be

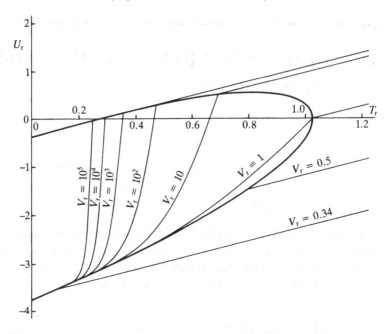

Figure 1.1. Internal energy U_r versus relative temperature $T_r = T/T_c$ (where the subscript c labels quantities at the critical point) for several fixed relative volumes $V_r = V/V_c$ for a van der Waals fluid, considered as an illustrative example. The relative internal energy is scaled to have the value 0 at the critical point; under the vapour pressure as $T_r \rightarrow 0$, $U_r(\text{liq}) \rightarrow (2\frac{1}{2} T_r - 3\frac{3}{4})$ and $U_r(\text{vap}) \rightarrow (1\frac{1}{2} T_r - \frac{3}{8})$. Temperature changes can be read off a horizontal line through and to the left of the point representing the initial state. I am grateful to Mr Erik Westerberg for writing the program which generated this figure.

expressed in terms of the fluid's equation of state, if known. Figure 1.1 shows the calculated results for a van der Waals fluid including both the single-phase and two-phase states. The free expansion is never used as a practical means of refrigeration and will not be discussed further here.

1.2.2   *Isentropic expansion or compression of a fluid*

The fluid is allowed to expand in a reversible and adiabatic way, performing external work in the process. One may visualise the fluid as being constrained in an insulated cylindrical container, one end of which is a frictionless piston whose motion against external pressure is kept to a very low speed to avoid irreversibility as far as possible. The process is

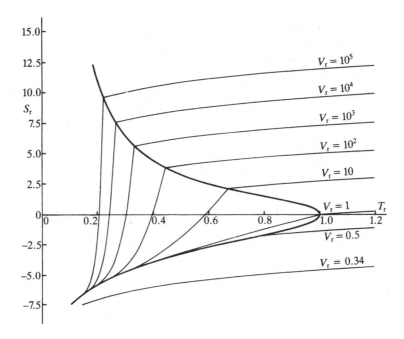

Figure 1.2. Entropy S_r versus relative temperature $T_r = T/T_c$ (where the subscript c labels quantities at the critical point) for several fixed relative volumes $V_r = V/V_c$ for a van der Waals fluid, considered as an illustrative example. The relative entropy is scaled to have the value 0 at the critical point (the van der Waals fluid does not obey the third law of thermodynamics); under the vapour pressure as $T_r \to 0$, S_r(liq) diverges towards minus infinity as $[2\frac{1}{2} \ln T_r - \ln(6\frac{3}{4})]$ and S_r(vap) diverges towards plus infinity as $[2\frac{1}{2} \ln T_r - \ln(6\frac{3}{4}) + 3\frac{3}{8}/T_r]$. Temperature changes can be read off a horizontal line through the point representing the initial state. I am grateful to Mr Erik Westerberg for writing the program which generated this figure.

quasistatic and in the single-phase state the temperature change can conveniently be expressed in the following form:

$$\left(\frac{\partial T}{\partial V}\right)_S = -\frac{T}{C_V}\left(\frac{\partial p}{\partial T}\right)_V. \tag{1.2}$$

This can be expressed in terms of the fluid's equation of state, if known, and for a monatomic perfect gas reduces to the familiar form

$$T_f = T_i(V_i/V_f)^{2/3}. \tag{1.3}$$

Figure 1.2 shows the calculated results for a van der Waals fluid including both the single-phase and two-phase states. This method, or a practical

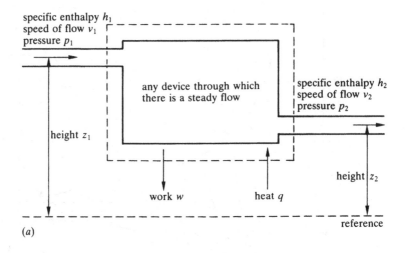

specific enthalpy h_1
speed of flow v_1
pressure p_1

any device through which there is a steady flow

specific enthalpy h_2
speed of flow v_2
pressure p_2

height z_1

work w

heat q

height z_2

reference

(a)

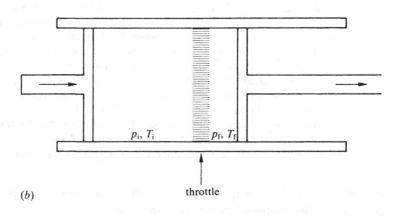

p_i, T_i p_f, T_f

(b)   throttle

Figure 1.3. (a) A general device through which there is a flow. The equation which relates the various quantities is

$$(h_2 - h_1) + \tfrac{1}{2}(v_2^2 - v_1^2) + g(z_2 - z_1) = q - w$$

where h_1, h_2, q, and w all refer to unit mass of flow. A special case of (a) is shown in (b) in which the device is a throttle whose effect is to render v_1, v_2, $(z_2 - z_1)$ negligible and in which q and w are zero. In this case the process is isenthalpic since $(h_2 - h_1) = 0$.

approximation to it, is used in low temperature refrigerators, most obviously in simple ³He and ⁴He 'pots'.

It is not practical to use a moving piston because of the large vapour volumes but the use of a vacuum pump to evaporate the liquid acts in a similar fashion. Also the basic principle of the Pomeranchuk refrigerator is recognisably the same, even though the two phases are in that case liquid and solid rather than vapour and liquid; moreover a movable piston *is* then used, usually in the form of bellows. See Chapter 4.

1.2.3 *Isenthalpic expansion of a fluid*

This takes place ideally in a device illustrated in texts on classical thermodynamics and in Figures 1.3(a) and 1.3(b). In Figure 1.3(b) the fluid is directed through a small orifice from a chamber maintained (by suitably controlled motion of a frictionless piston) at pressure p_1 into a second chamber maintained (by suitably controlled motion of another frictionless piston) at pressure p_2. All the walls are insulating and $p_1 \geqslant p_2$. It is trivial to show that the enthalpy H remains constant. The process is

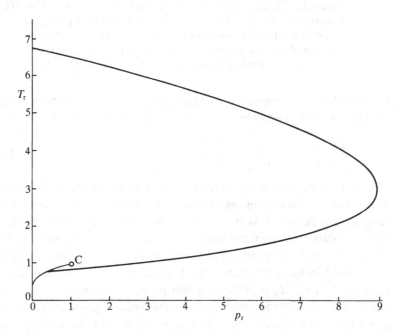

Figure 1.4. The inversion curve of a van der Waals fluid, considered as an illustrative example.

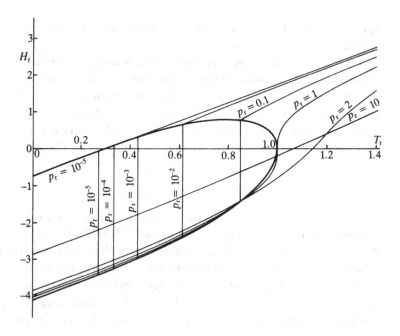

Figure 1.5. Relative enthalpy versus relative temperature $T_r = T/T_c$ (where the subscript c labels quantities at the critical point) for several fixed relative pressures $p_r = p/p_c$ for a van der Waals fluid, considered as an illustrative example. I am grateful to Mr Erik Westerberg for writing the program which generated this figure.

irreversible and entropy rises. In the single-phase state the temperature change can conveniently be expressed in the following form:

$$\left(\frac{\partial T}{\partial p}\right)_H = + \frac{1}{C_p}\left[T\left(\frac{\partial V}{\partial T}\right)_p - V\right]. \tag{1.4}$$

This is in general not zero (except for a perfect gas) and may be positive or negative depending on the pressures and initial temperature. In low temperature applications one is normally operating within the fluid's inversion curve and then the result is cooling. Figures 1.4 and 1.5 show respectively the inversion curve and calculated results for a van der Waals fluid including both the single-phase and two-phase states. In Figure 1.4 $T_r = T/T_c$ is plotted against $p_r = p/p_c$ where the subscript c labels quantities at the critical point. The critical point at $T_r = 1$ and $p_r = 1$ is labelled C in the figure and the curve which links it to the origin at $T_r = 0$ and $p_r = 0$ is the vapour pressure curve. All points in this figure, except those which lie on the vapour pressure curve, refer to the single-phase state. All points within the enclosed area have a positive value of the quantity specified in equation (1.4), that is, cooling can be achieved by an isenthalpic process in which the fluid passes from a higher to a lower pressure. The maximum inversion temperature for a van der Waals fluid is exactly $T_r = 6\frac{3}{4}$ and the tip of the curve is exactly at $T_r = 3$ and $p_r = 9$.

In Figure 1.5 enthalpy is plotted against relative temperature for several fixed relative pressures. The relative enthalpy is scaled to have the value 0 at the critical point; under the vapour pressure as $T_r \to 0$, $H_r(\text{liq}) \to (2\frac{1}{2} T_r - 4\frac{1}{8})$ and $H_r(\text{vap}) \to (2\frac{1}{2} T_r - \frac{3}{4})$. Temperature changes can be read off a horizontal line through and to the left of the point representing the initial state.

This isenthalpic method, or a practical approximation to it, is used in low temperature refrigerators in that cooling and liquefaction can be achieved by forcing the fluid through a constriction into a lower pressure chamber. The ideal text-book case ignores changes in potential and kinetic energies between the two sides of the constriction; it is not difficult to incorporate these into the formulation but for the purposes of this book they can safely be neglected. In the cases of simple evaporation 'pots', in which the pressure difference across the liquid/vapour interface is zero, and the chemical potential is the same on both sides, one may equivalently describe the process (evaporation or condensation) as isentropic and as isenthalpic. Finally, real low-temperature refrigerators almost never have moving pistons (the exception mentioned above is the Pomeranchuk refrigerator) but vacuum pumps perform equivalent functions.

1.2.4   Adiabatic (isentropic) demagnetisation of a paramagnet

In this case one thinks of a solid array, not necessarily in the form of an ordered crystal, of atomic or nuclear magnetic moments. The most frequently used materials include CMN (cerous magnesium nitrate, an electronic paramagnet), copper (a nuclear paramagnet), and $PrNi_5$ (praesodymium nickel five, an intermetallic compound which is a hyperfine-enhanced nuclear paramagnet). Figure 1.6 shows the entropy of CMN as a function of temperature for several applied magnetic fields and illustrates the principle; for example, a demagnetisation from (A) 0.1 tesla at an initial temperature of 100 mK to (B) zero field reduces the temperature to about 3 mK.

The attainment of the initial state would normally require a fairly simple

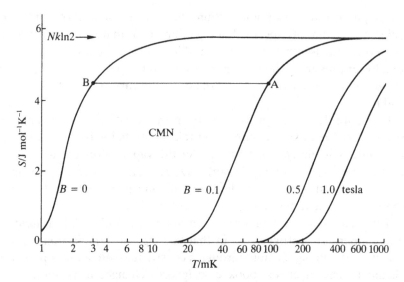

Figure 1.6. The entropy of crystalline CMN versus temperature for several applied magnetic fields directed along the crystallographic *a*-axis. From data given by Fisher *et al.* (1973).

dilution refrigerator (see Chapters 2 and 3) and some form of heat switch to remove the heat of magnetisation which appears when the refrigerant is magnetised. When the demagnetisation is made, the heat switch has to be open to minimise heat leaks. The full cycle of magnetisation and demagnetisation is indicated in Figure 1.7, and more detail is given in Chapter 5.

The thermodynamics of the processes can conveniently be summarised in two equations. First, the heat of magnetisation released when the applied field is increased from zero to B_i at T_i is (assuming reversibility):

$$Q_{mag} = T_i[S(0, T_i) - S(B_i, T_i)]. \tag{1.5}$$

Second, the amount of cooling can be calculated in general by integrating the following equation:

$$\left(\frac{\partial T}{\partial B}\right)_S = -\left(\frac{\partial T}{\delta S}\right)_B \left(\frac{\partial S}{\partial B}\right)_T \tag{1.6}$$

$$= -\frac{TV}{C_B}\left(\frac{\partial M}{\partial T}\right)_B \tag{1.7}$$

where C_B/V is the specific heat per unit volume at constant field B, and M is the magnetisation (that is, the magnetic moment per unit volume). For

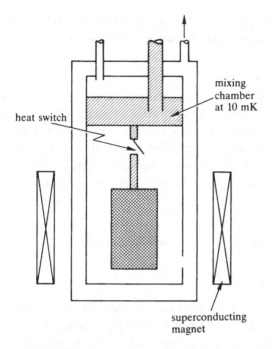

Figure 1.7. Diagram illustrating the full cycle of magnetisation and demagnetisation. The initial state may be taken as $T_i = 10$ mK (achievable by contact through a closed heat switch with the mixing chamber of a dilution refrigerator) with $B = 0$. The steps are as follows. (i) The field is raised from zero to B_i (typically using a superconducting solenoid), and the heat of magnetisation is conducted through the heat switch to the mixing chamber. (ii) The heat switch is opened so that the refrigerant is thermally isolated. (iii) The field is reduced slowly (ideally isentropically) to B_f and the temperature falls to T_f. The initial state is easily regained if required by closing the heat switch and reducing the field to zero.

a perfect paramagnet, that is, one in which interactions between the atomic or nuclear moments are ignored, it is a standard exercise in statistical mechanics to show that the entropy of N moments γ with quantum number J the entropy is given by

$$\frac{S}{Nk_B} = + \left[\ln \sinh \left(\frac{2J+1}{2J} \frac{\gamma B}{k_B T} \right) - \ln \sinh \left(\frac{1}{2J} \frac{\gamma B}{k_B T} \right) \right]$$

$$- \frac{\gamma B}{k_B T} \left[\left(\frac{2J+1}{2J} \right) \coth \left(\frac{2J+1}{2J} \frac{\gamma B}{k_B T} \right) - \left(\frac{1}{2J} \right) \coth \left(\frac{1}{2J} \frac{\gamma B}{k_B T} \right) \right]. \quad (1.8)$$

For the perfect paramagnet it can be seen by inspection of equation (1.8) that if B/T is constant then the process is isentropic, and vice versa. It follows that

$$T_f = (B_f/B_i) T_i. \tag{1.9}$$

This form is often adequate but it cannot of course be accepted in the limiting case of demagnetisation to $B_r = 0$ because then the assumption of negligible interactions (compared with $k_B T$) between the moments breaks down. It is generally safer to think of the effective field B_{eff} on a moment being a combination of the applied field B and an internal field B_{int} characteristic of the actual paramagnet in use. A suitable combination is to take

$$B_{eff}^2 = B^2 + B_{int}^2 \tag{1.10}$$

and this leads to an improved version of equation (1.9):

$$T_f = \left(\frac{B_f^2 + B_{int}^2}{B_i^2 + B_{int}^2} \right)^{1/2} T_i. \tag{1.11}$$

This correctly places $T_f = 0$ out of reach, consistently with the third law of thermodynamics in that even when $B_f = 0$ the predicted final temperature is not zero, and when in addition $B_i \gg B_{int}$ it reduces to the acceptable limit

$$T_f = (B_{int}/B_i) T_i. \tag{1.12}$$

1.3   Thermometry

1.3.1   The Kelvin scale

The existence of temperature as a function of state is a direct consequence of the zeroth law of thermodynamics. At temperatures which are not too close to the absolute zero, the constant-volume gas thermometer serves as a fundamental thermometric device. According to this,

$$T = 273.16 \lim_{p_t \to 0} \left(\frac{p}{p_t} \right), \tag{1.13}$$

where p is the gas pressure at temperature T, and p_t is the gas pressure at the triple-point temperature of water, defined as 273.16 K. This definition is found to provide a temperature scale which is independent of the properties of the particular gas used, but which of course depends on the properties of gases in general. Thus equation (1.13) cannot be used far below 1 K, simply because no uncondensed gases are available. It is

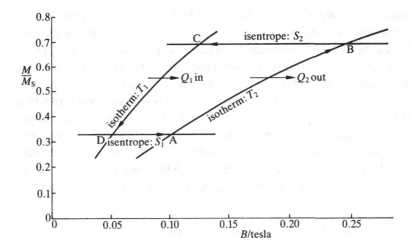

Figure 1.8. Fractional magnetisation versus applied field for an ideal paramagnetic material, showing a reversible Carnot cycle ABCD formed by two isotherms (at T_1 and T_2) and two isentropics (S_1 and S_2).

necessary therefore to move beyond equation (1.13) to an even more fundamental position based on the second law of thermodynamics. We imagine a Carnot cycle to be run between two reservoirs, one at the temperature T to be measured and the other to be at the temperature T_t of the triple point of water, defined to be 273.16 K. Now the Carnot cycle is reversible and so, from the second law, we can assert that

$$\oint \frac{\mathrm{d}Q}{T} = 0 \qquad \text{(reversible process)} \qquad (1.14)$$

which reduces to the form

$$T = 273.16 \left(\frac{Q}{Q_t} \right). \qquad (1.15)$$

It is straightforward to show that equation (1.13) is merely a special case of equation (1.15), which has the massive advantage that it remains true *irrespectively of the choice of thermometric parameter*. A magnetic Carnot cycle is illustrated in Figure 1.8 but any Carnot cycle will do.

We are thus free to choose a method suitable for use in the millikelvin range of temperatures and to link that method directly to the Kelvin scale. This linking may not be easy to achieve in practice but it is important to retain confidence in the notion that a temperature measured in one way in Washington or Beijing or Helsinki can be correlated with a temperature measured in another way in Paris or Moscow or London. The existence of national standards laboratories around the world bears witness to the importance that physicists attach to these considerations. One example will suffice to illustrate this point. There has been much debate about the possible existence and origin of correction terms to be added to the expression $C = \gamma T$ for the low temperature heat capacity of normal liquid ^{3}He. If the terms really exist, then they require a theoretical interpretation; if on the other hand they are merely artifacts of faulty thermometry, then efforts have to be made in other directions. It is crucial to be able to distinguish between these two possibilities.

1.3.2   International practical temperature scales

Most researchers prefer to depend on the work of specialists to link practical thermometry to the Kelvin scale, although this is not always possible, particularly in the most careful work in the millikelvin range of temperatures. Two examples must suffice. First, in using the vapour pressure of liquid ^{3}He down to about 250 mK, it is universal practice to turn to agreed tables of p_{vap} as a function of T (see Chapter 6). The compilation of such tables, and their correlation with the true Kelvin scale represents activity of considerable magnitude which is subsequently taken on trust by users. Second, it has been common in using the magnetic susceptibility χ of CMN as a thermometric parameter to use a powdered sample of cylindrical geometry with diameter equal to length. From a theoretical point of view it would be preferable to use prolate spheroid geometry, but this is less practical to achieve. So much work has had to be done to relate χ to T and tables are then subsequently used as references; this is particularly important below about 10 mK where corrections to Curie's law ($\chi \propto T^{-1}$) become significant. More details will be given in Chapter 6, but the important point to note here is that international agreements about practical thermometry have to exist. At the time of writing (July 1987) the current agreement is the International Practical Temperature Scale of 1968 ([IPTS–68, see alphabetic references under Comité International des Poids et Mesures (1969) and under National Physical Laboratory (1976)). IPTS-68 is due to be replaced quite soon with a scale designed to be closer to the true Kelvin scale. As it stands, IPTS-68 is not of much direct concern for millikelvin technology because the lower end of its temperature range is 13.81 K although there is an appendix devoted to the use of ^{3}He and ^{4}He vapour pressures between 0.2 and 5.2 K. This appendix has to a large extent been superseded by the 1976 Provisional 0.5–30 K Temperature Scale (EPT-76, see alphabetic references under Bureau International des Poids et Mesures (1979); see also Durieux and Rusby (1983) who give ^{3}He vapour pressures down to 200 mK). Ultimately of course there will have to be agreed scales to much lower temperatures, perhaps using such fixed points as the zero-field superconducting transitions of $AuIn_2$, $AuAl_2$, Ir, Be and W (respectively $T_c \approx 207, 161, 99, 23$ and 15 mK) and the melting-pressure and zero-pressure superfluid transitions of ^{3}He (respectively $T_c \approx 3$ and 1 mK). Meanwhile researchers must choose their authorities with care and publish enough detail to enable sub-sequent temperature translations to be made unambiguously if necessary.

2 Properties of fluid ^{3}He/^{4}He mixtures

2.1 Preamble

The dilution refrigerator, to be dealt with in Chapter 3, can only be fully understood in terms of the properties of the two helium isotopes ^{3}He and ^{4}He and their mixtures. The purpose of this chapter is very briefly to review the relevant properties indicating, where necessary, sources giving more details. A useful general reference is by Wilks and Betts (1987) which also serves as an introduction to the research literature. As in the whole of this book, the reader is expected to derive at least as much information from the figures as from the minimal text.

2.2 Phase diagrams

The phase diagrams of ^{4}He, ^{3}He, and their mixtures are given below as Figures 2.1 and 2.2 respectively. Particular points of interest are discussed in the rubric and text following each figure.

Liquid ^{4}He can be solidified by pressures above about 25 bar at temperatures below about 1 K. Higher pressures are needed at higher temperatures but below 1 K the melting pressure is nearly constant. The liquid phase is superfluid when the pressure is below the melting pressure and when in addition the temperature is below the lambda line (labelled in the figure) which runs from 2.17 K at $p = 0$ bar to 1.76 K at $p = p_{\text{melt}} \approx 30$ bar. The superfluid phase is called He-II to distinguish it from normal He-I.

Liquid ^{3}He has a pronounced minimum in its melting pressure of $p_{\text{melt}} \approx 29$ bar at $T \approx 0.32$ K and at lower temperatures p_{melt} rises to about 34 bar as $T \rightarrow 0$ K. Normal (that is, non-superfluid) liquid ^{3}He persists down to a few millikelvin where it is well-described as a Fermi liquid. At lower temperatures, the liquid becomes superfluid at T_c (analogous to the lambda temperature T_λ for ^{4}He). T_c runs from 1.0 mK

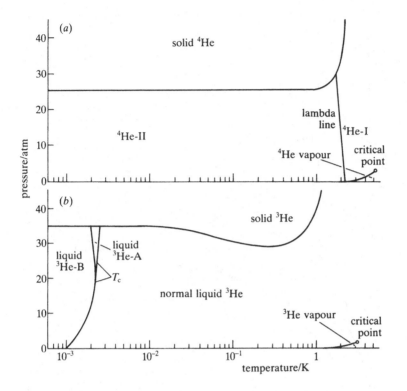

Figure 2.1. The phase diagram of ^{4}He and ^{3}He, shown with the same scales for convenient comparison (from McClintock *et al.* (1984)). The temperature axis has a logarithmic scale.

at $p = 0$ bar to 2.8 mK at $p = p_{\text{melt}} \approx 34$ bar as shown in the figure. At T_c the liquid in zero magnetic field enters the ^{3}He-A phase for pressures greater than 21.5 bar or the ^{3}He-B phase for pressures less than 21.5 bar. Both ^{3}He-A and ^{3}He-B are superfluid.

Next we need the phase diagram of liquid ^{3}He/^{4}He mixtures and this is shown in Figure 2.2. It is particularly important for the understanding of dilution refrigerators. Mixtures with ^{3}He concentration X_3 greater than about 6% will always separate into two phases if cooled sufficiently far below 0.87 K; at temperatures higher than 0.87 K separation is never observed for any concentration. The two-phase region is bounded, as shown in the figure, by (i) the $T = 0$ K axis, (ii) a line running from $T = 0$ K and $X \approx 6\%$ to the tricritical point at $T = 0.87$ K and $X_3 = 67\%$, and (iii) a line running from the tricritical point to $T = 0$ K and $X_3 = 100\%$. At both points where the phase separation line approaches

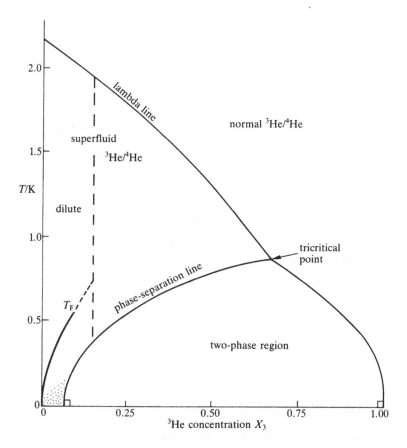

Figure 2.2. Phase diagram of liquid ^{3}He/^{4}He mixtures at zero pressure. (The diagram is from Wilks and Betts (1987); it is based on data from Laheurte and Keyston (1971), Ghozlan and Varoquaux (1975), Ebner and Edwards (1970), Dokoupil *et al.* (1959), de Bruyn Ouboter *et al.* (1960), and Alvesalo *et al.* (1971).) This diagram relates only to the liquid phase and its axes are temperature and concentration.

the $T = 0$ K axis, it does so along a perpendicular, consistently with the third law of thermodynamics. The liquid mixtures are superfluid only if they are in states which lie on the low-concentration side of the lambda line which runs from $T = 2.17$ K and $X_3 = 0\%$ (that is, pure ^{4}He) to the tricritical point. When considering the behaviour of dilute mixtures (roughly those with $X_3 < 15\%$) it is useful and justifiable to think of the ^{3}He atoms as constituting a 'gas-like' solute; this 'gas' may be nearly classical or nearly fully Fermi-degenerate or somewhere in between, depending on whether $T \gg T_f$, $T \ll T_f$, or $T \approx T_f$. The line of $T_f(X_3)$ is shown in the figure.

Within the two-phase region, the ^{3}He-rich phase (RHS) floats on the ^{4}He-rich phase (LHS) as $T \to 0$ (say, $T \leqslant 200$ mK) and the following approximations may be used:

$$\text{RHS: } X_4 \equiv (1 - X_3) = 0.85 \ T^{3/2} \exp(-0.56/T) \tag{2.1}$$

so that

$$\frac{dX_3}{dT} \to -\infty \quad \text{as} \quad T \to 0 \quad \text{(third law of thermodynamics)} \tag{2.2}$$

[see Laheurte and Keyston (1971)].

$$\text{LHS: } X_3 = 0.0648 \ (1 + 8.4 \ T^2 + 9.4 \ T^3) \tag{2.3}$$

so that

$$\frac{dX_3}{dT} \to +\infty \quad \text{as} \quad T \to 0 \quad \text{(third law of thermodynamics)} \tag{2.4}$$

[see Ghozlan and Varoquaux (1975), and Edwards *et al.* (1969)].

2.3 Dilute mixtures (that is, $X_3 \lesssim 0.15$)

When $T \leqslant 500$ mK, the solvent ^{4}He has (to a good approximation) zero viscosity, zero entropy and therefore zero specific heat. It is often described as a 'massive vacuum'. The consequence is that the dissolved ^{3}He behaves rather like a perfect gas, but with an effective mass m_3^* approximately equal to $2.4 m_3$. A very useful paper on the thermodynamic properties of dilute mixtures has been given by Kuerten *et al.* (1985).

2.4 Fermi degeneracy of solute helium-3

From the standard statistical mechanics of a perfect Fermi gas,

$$T_f(X_3) = \frac{h^2}{8 k_B m_3^*(X_3)} \left(\frac{3NX_3}{V}\right)^{2/3}. \tag{2.5}$$

Very roughly, if $T < \frac{1}{3} T_f$, the behaviour should approximate that of a Fermi degenerate gas, but if $T > T_f$ it should approximate that of a classical gas. Both limiting cases are observed, for example, see the specific heat in Figures 2.3 and 2.4.

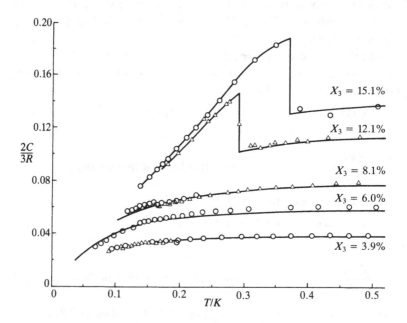

Figure 2.3. Specific heat of mixtures above 70 mK, plotted as $\frac{2}{3}C/R$ versus T. Various concentrations X_3 are represented. The peaks shown with higher concentrations are due to phase separation. At the higher temperatures $\frac{2}{3}C/R$ approaches X_3, corresponding to classical perfect gas-like behaviour $(C = \frac{3}{2}X_3R)$. (Ref: Edwards *et al.* (1965)).

2.5 Mixtures and the two-fluid model

In pure ^{4}He-II the two-fluid model (with normal component denoted by subscript n, and superfluid component denoted by subscript s) may be summarised in the following points: (i) $\varrho_n + \varrho_s = \varrho$, (ii) $\eta_s = 0$ but $\eta_n \neq 0$, (iii) entropy is carried only by normal component which consists of excitations of two distinct types – phonons and rotons). This model works well. Added ^{3}He becomes part of the normal component, rather like another type of excitation but one which persists as $T\rightarrow 0$ K leaving a finite ϱ_n (in ^{4}He-II, $\varrho_n\rightarrow 0$ as $T\rightarrow 0$ K).

2.6 Osmotic pressure

Figure 2.5 is a visual aid to reinforce the notion that osmotic pressure is a very real effect which can produce substantial level differences. A

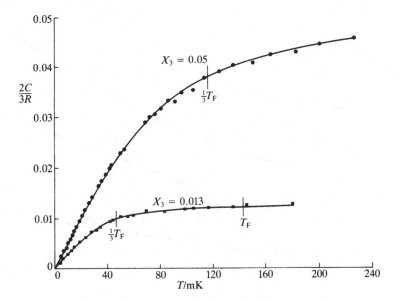

Figure 2.4. Specific heat of mixtures below 230 mK, plotted as $\frac{2}{3}C/R$. Two concentrations X_3 are represented. The changeover from classical perfect gas-like behaviour at higher T to the linear behaviour of degenerate Fermi gas-like behaviour at lower T is seen to occur in the region of $T = \frac{1}{3}T_f$ (see Anderson *et al.* (1966)).

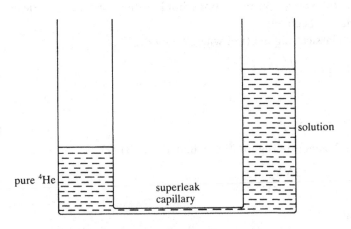

Figure 2.5. Osmotic pressure (see text).

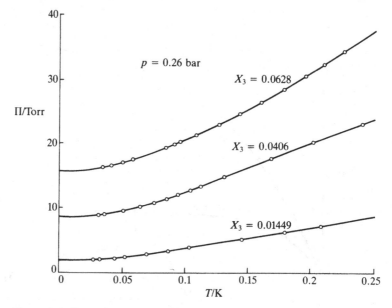

Figure 2.6. Osmotic pressures of some dilute mixtures at a pressure of 0.26 bar. Note that the pressure axis is labelled in Torr, so the pressures are quite large. (Ref: Landau *et al.* (1970)).

superleak is an almost perfect semi-permeable membrane which is able to prevent ^{3}He atoms passing from the solution to the pure liquid ^{4}He-II while allowing free passage to the superfluid component. Osmotic pressures (Π) can be quite large (for example, as $T \to 0$ for $X_3 = 0.01$, Π is about 20 cm of liquid helium). See Figure 2.6 above.

The following approximations hold in the classical and degenerate régimes respectively.

In the classical régime (i.e. when $T \geq T_f$),

$$\Pi \approx \frac{NX_3 k_B T}{V}, \qquad (2.6)$$

so that

$$\Pi \propto X_3 T. \qquad (2.7)$$

In the degenerate régime (i.e. when $T \leq (1/3) T_f$),

$$\Pi \approx \frac{2NX_3 k_B T_f}{5V}, \qquad (2.8)$$

so that

$$\Pi \propto X_3^{5/3}. \qquad (2.9)$$

2.7 Vapour pressure

This is a function of X_3 and T. As $T \to 0$, $p_{\text{vap}} \to 0$. As we shall see, the chief interest for the purposes of dilution refrigeration is in concentrations in the range 0.01–0.05 and temperatures in the range 0.5–1.0 K. In those ranges, one may use the following approximations [see Vvedenskii and Peshkov (1972)]:

$$P = P_4 + P_3 \tag{2.10}$$

where

$$P_4 = P_4^0 (1 - X_{3l}) \tag{2.11}$$

and

$$P_3 = a P_3^0 X_{3l} \tag{2.12}$$

where

$$a = 4.75(1/T - 0.17). \tag{2.13}$$

X_{3l} is the concentration in the liquid, and P_4^0 and P_3^0 are the vapour pressures of the pure isotopes. For example, at 0.7 K with $X_{3l} = 0.01$ we find

$$P_4^0 = 000.292 \text{ Pa} \tag{2.14}$$
$$P_3^0 = 179.876 \text{ Pa} \tag{2.15}$$

giving

$$P_4 = 000.289 \text{ Pa} \tag{2.16}$$

and

$$P_3 = 010.753 \text{ Pa} \tag{2.17}$$

where 1 Pa $\equiv$ 7.501 mTorr. So $P_3 + P_4 = 11$ Pa and $P_3/P_4 = 37$.

2.8 Transport properties

Heat flow in solutions is affected by scattering processes involving phonons and rotons and by the heat flush effect which sweeps ^{3}He atoms along with the heat current to produce a concentration gradient in the liquid. However, at sufficiently low temperature both these effects are unimportant because the numbers of excitations become negligible, and we may apply Fermi gas theory just to the 'gas' of ^{3}He atoms. If the temperature is also low enough for this 'gas' to be degenerate then we expect that the mean free path varies as $(T_f/T)^2$, and this leads to a crude prediction that the thermal conductivity should be proportional to X_3/T. There is fair agreement with experiments with regard to the temperature

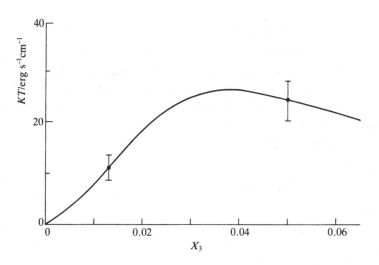

Figure 2.7. Thermal conductivity of mixtures, plotted as the product KT of the thermal conductivity and the temperature versus the concentration X_3. The data points are due to Abel *et al.* (1967) and the theoretical curve is due to Wei-Chan Hsu and Pines (1985). The theory is based on a choice of assumed interaction potential between ^{3}He atoms, the choice being made in such a way as to optimise agreement with as many different experiments as possible.

dependence. For example, with $X_3 = 0.013$, measurements show $KT = 11$ erg s^{-1} cm^{-1} (Abel *et al.* (1967)). But the X_3 dependence for the thermal conductivity (as for the other transport properties) needs more experimental evidence, theory being ahead of experiment at present. This is illustrated in Figure 2.7 which shows recent detailed theoretical prediction (curved line): the crude degenerate Fermi gas model would give a straight line through the origin and existing data is not extensive enough to resolve the difference.

A similar crude treatment suggests that the viscosity will be proportional to $(X_3)^{5/3}/T^2$. This has not so far been fully satisfactorily demonstrated and may not in fact be in accordance with observations, perhaps because of surface effects. Figure 2.8 shows some recent results. The lowest set of raw data are for $X_3 = 1.07\%$ (open circles show corrected data). The next set up are for $X_3 = 5.2\%$ (filled circles show corrected data). The top set are for $X_3 = 99.95\%$, almost pure. The corrections were for mean free path and surface scattering effects. This property is not easy to measure and is notorious for appearing to depend on frequency (for methods which employ oscillation) and on surface conditions.

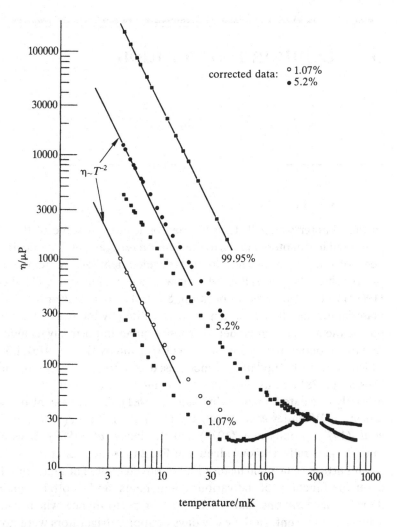

Figure 2.8. Viscosity of mixtures. (Ref: Ritchie (1985).)

3   Dilution refrigeration

3.1   Preamble

In this chapter we shall describe the application of some of the ideas presented in Chapters 1 and 2 to the dilution refrigerator. This device has been of crucial importance for the development of low temperature physics since the mid-sixties when it was first demonstrated (Hall *et al.* (1966)) to be a practical proposition. Several sources have provided the material for this chapter, and may be consulted by the reader wishing to pursue the matter in great depth. These sources importantly include two research monographs by Lounasmaa (1974) and by Betts (1976), both of which need to be updated, a much-used conference review article by Frossati (1978) and its sequel by Vermeulen and Frossati (1987), and a privately-circulated manual by Sagan (1981). There are also useful articles by Wheatley *et al.* (1968) and (1971), Niinikoski (1976), Frossati *et al.* (1977), Lounasmaa (1979) and Bradley *et al.* (1982). It would I think be generally acknowledged that Frossati is the master in this field, and his article is the main inspiration for this chapter. The aim is briefly to cover the functions of the various components, and to offer a guide to design considerations and to the sort of performance which can be achieved at present. In the early days dilution refrigerators were home-made but most users now buy them commercially in much the same way as consumers buy domestic refrigerators, though at considerably greater expense. The main supplier at present is the Oxford Instrument Company based in the UK.

3.2   Evaporation cooling

It is useful to begin by having in mind a simple image of an evaporation cooler as shown in Figure 3.1.

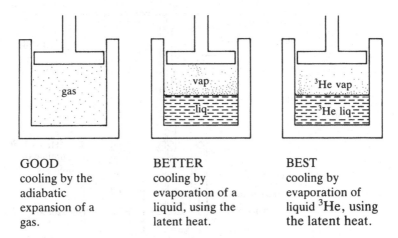

GOOD
cooling by the
adiabatic
expansion of a
gas.

BETTER
cooling by
evaporation of a
liquid, using the
latent heat.

BEST
cooling by
evaporation of
liquid ^{3}He, using
the latent heat.

Figure 3.1. Cooling by expansion of fluids.

The thermodynamic principles of the simple evaporation cryostat or 'pot' are outlined in Chapter 1. ^{3}He is the best choice of simple refrigerant for the attainment of the lowest possible temperatures because it has the highest vapour pressure. The dilution refrigerator (this chapter) and the Pomeranchuk refrigerator (Chapter 4) are sophisticated variants of the general idea. They are more successful, if more elaborate, and also make use of the unusual properties of ^{3}He.

^{3}He cryostats may be 'single-shot' (i.e. the pumped ^{3}He gas is not returned) or 'recirculating'. Figure 3.2 is a schematic representation of a possible layout for a recirculating ^{3}He cryostat. Only the basic components are shown, and the main ^{4}He bath has been stylised to avoid complicating the diagram. In reality the 4.2 K bath usually surrounds the outer vacuum jacket. Sometimes the main bath itself is pumped, thus avoiding the need for a subsidiary '^{4}He pot'. The constriction where the ^{4}He pumping line enters the ^{4}He pot is to reduce film flow and thus enhance pumping efficiency. The ^{3}He pot is usually provided with some extra exchange area to aid thermal contact and to reduce gradients. ^{3}He cryostats may be designed according to needs to operate down to about 300 mK and up to 1 K or more.

3.3 Layout of components in a dilution refrigerator

We shall begin by working through a progression of figures towards the semi-realistic representation of Figure 3.6. Firstly it is helpful to

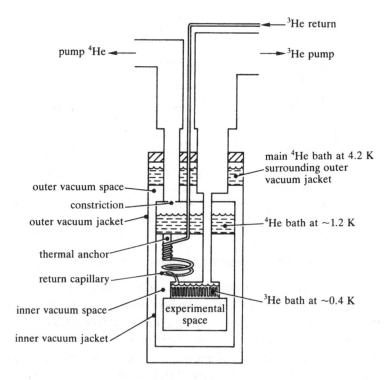

Figure 3.2. A recirculating ³He cryostat or 'pot'. For more details, see Rose-Innes (1973), Lounasmaa (1974), Betts (1976), or White (1979).

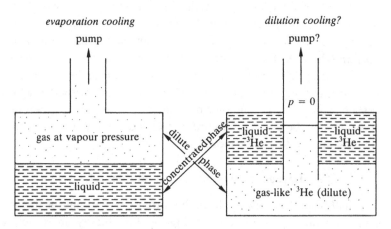

Figure 3.3. Understanding dilution refrigeration (see text).

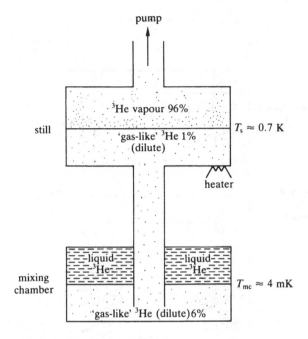

Figure 3.4. A better model.

recognise the coldest part, the 'mixing chamber', as a sort of inverted evaporation 'pot' as seen in Figure 3.3. On the left is a simple evaporation cooler using ^{3}He at, say, 400 mK where the vapour pressure is 3.6 Pa and so easily pumpable. On the right is a vessel containing a phase-separated mixture of ^{3}He and ^{4}He (see Figure 2.2) at, say 4 mK. The upper phase is almost pure ^{3}He and the lower phase contains about 6% of ^{3}He which may (see Chapter 2) be regarded in several ways as 'gas-like' with the ^{3}He atoms moving freely in the superfluid 'massive vacuum' provided by the solvent ^{4}He-II. If only it were possible to pump ^{3}He out of the lower phase then there would be a net flow of ^{3}He downwards across the phase interface and cooling would be achieved as on the left hand side. Unfortunately the vapour pressure at 4 mK is effectively zero and no vacuum pump could achieve the desired result.

In order to allow pumping, an extra feature (the still) must be included, as shown diagrammatically in Figure 3.4 in which a very slightly more realistic dilution refrigerator is shown in which the ^{3}He atoms are removed by pumping on a still (maintained typically at about 0.7 K by an electrical heater) with an ordinary vacuum pump at room temperature. The result of this pumping is that atoms are drawn downwards across the

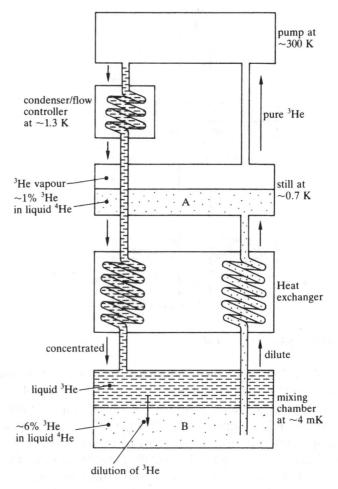

Figure 3.5. Layout of components in a conventional dilution refrigerator diagrammatic form.

interface from the concentrated to the dilute phase in the mixing chamber, producing cooling through a latent heat of mixing. The temperatures and concentrations shown are consistent with each other for a steady state.

The complete device is usually more complicated because it is desirable to circulate ^{3}He continuously and this means including effective heat exchangers (see Figure 3.5). When the refrigerator is operating, almost pure ^{3}He arrives in the upper phase in the mixing chamber, then to the lower ^{4}He-rich phase, from which it is distilled, and then finally returned

Table 3.1 *Properties of pure and saturated dilute ^3He*

Property	Pure ($X_3 = 1$) concentrated. C = concentrated	Saturated and dilute ($X_3 = 0.0648$). D = dilute
Specific heat = specific entropy	$25\ T$ J mol^{-1} K^{-1}	$107\ T$ J mol^{-1} K^{-1}
Enthalpy	$12.5\ T^2$ J mol^{-1} K^{-2}	$94.5\ T^2$ mol^{-1} K^{-2}
^{3}He molar volume	36.86 cm^3	424.4 cm^3
Viscosity	$2.2\ T^{-2}$ μP	$0.3\ T^{-2}$ μP
Thermal conductivity	$3.3\ T^{-1}$ μJ cm^{-1} s^{-1} deg^{-1}	$2.4\ T^{-1}$ μJ cm^{-1} s^{-1} deg^{-1}
Osmotic pressure relative to C	0 bar	$22.4 + 1.0 \times 10^3\ T$ $- 1.2 \times 10^3\ T^4$ mbar

by a different route to the ^{3}He rich phase. The cooling is produced in the mixing chamber by the expansion (more accurately, dilution) of the 'gas' of ^{3}He atoms as they pass from the ^{3}He rich phase into the dilute condition of the ^{4}He rich phase. The heat extraction rate depends on the molar rate of ^{3}He flow across the interface in the mixing chamber, $\dot{n}_3$, and on the enthalpy difference between the condensed and dilute states of the ^{3}He. As both these states are Fermi-degenerate, the enthalpy difference is proportional to T^2 and numerically the heat extraction rate is 82 $\dot{n}_3 T^2$. The circulation of ^{3}He is maintained by distilling almost pure ^{3}He from the 1% solution in the 'still' (Figure 3.4) at about 0.7 K. The magnitude of $\dot{n}_3$ is determined by the operating conditions and the speed of the pumping system, and is typically 10^{-4}–10^{-3} mol s^{-1}, and in order to maintain a given rate $\dot{n}_3$ it is necessary to supply heat to the still at a rate proportional to $\dot{n}_3$ (numerically about 40 $\dot{n}_3$ W). The ^{3}He lost from the still by distillation is steadily replaced by the passage of ^{3}He atoms across the phase interface in the mixing chamber and then upwards through the heat exchangers to the still. The pumped ^{3}He is condensed on the high pressure side of the pump and passes down through the heat exchangers where it is cooled before arriving in the mixing chamber, completing a cycle. The lowest temperature for a particular refrigerator is reached when the heat extraction rate is just balanced by stray heat leaks and by any inefficiency in the heat exchangers. In fact the correct design of these components is crucial to the successful operation of the refrigerator. Many square metres of contact area are needed to overcome the Kapitza boundary resistance (we shall discuss this in greater

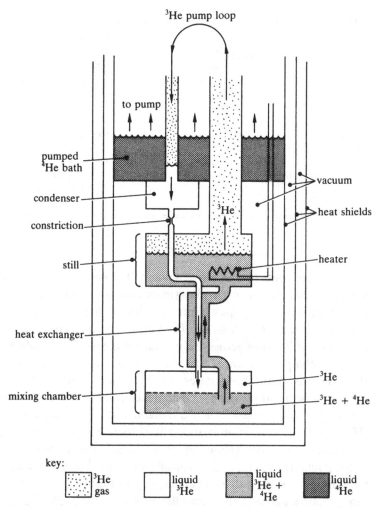

Figure 3.6. A semi-realistic diagram of a dilution refrigerator.

detail below) and the exchangers are usually produced by sintering fine metallic powders (often silver) to make porous blocks through which the helium can pass. See Figure 3.6. The ideal heat extraction rate assuming one-shot operation or perfect heat exchange, and no heat load, is

$$\dot{n}_3 T(S_D - S_C) = \dot{n}_3 (H_D - H_D) = 82 \, \dot{n}_3 T^2 \text{ W.} \tag{3.1}$$

3.4 Startup (Frossati (1978))

The circuit is filled with a mixture of ^{3}He and ^{4}He which is precooled to 4 K in a liquid helium bath and then condensed to about 1.3 K in a pumped ^{4}He pot which may be recirculating. The liquid mixture is homogeneous at this stage. There should be a free surface in the still; when this is pumped, the mixture cools and eventually phase-separates. At first there is only dilute mixture in the mixing chamber and concentrated ^{3}He floating in the still, but this is soon pumped away and injected into the mixing chamber and the phase interface is then correctly located. But the temperature falls only to about 0.3 K because the vapour pressure in the still is too low. The still heater is now switched on to raise its temperature in order to stimulate circulation. Typically $T_2 \approx 0.7$ K and $\dot{Q}_s \approx 40\dot{n}_3$ W and we return to the factors governing this choice below.

The dilution process now continues and cooling proceeds until the heat extraction rate 82 $\dot{n}_3 T^2$ is balanced by the sum of the various heat inputs which may be present.

3.5 Amount and concentration of mixture

Malfunction will certainly result if in the supply gas either the total molar amount ($n_3 + n_4$) or the molar isotopic ratio (n_3/n_4) lie outside acceptable limits. The two determining criteria for these limits are that the free surface be correctly positioned in the still and the phase interface be correctly positioned in the mixing chamber, so the internal geometry of the refrigerator needs to be known with reasonable accuracy.

3.6 The still

Firstly, the concentration in the still depends on temperature and is determined by the balancing of osmotic pressures of ^{3}He in the still and mixing chamber (see Figure 3.7). In the still (with $T = T_s$ and $X_3 = X_{3s}$) the dissolved ^{3}He 'gas' is classical ($T_s \gg T_f$) and its osmotic pressure is given by

$$\Pi = \frac{N X_{3s} k_B T_s}{V}. \tag{3.2}$$

In the mixing chamber (with $T = T_{mc}$ and $X_3 = X_{3mc}$) the dissolved 'gas' is Fermi degenerate ($T_{mc} \ll T_f$, given by equation (2.5)) and its osmotic pressure is given by

$$\Pi = \frac{2 N X_{3mc} k_B T_f(X_{3mc})}{5V}. \tag{3.3}$$

Equating (3.2) and (3.3) gives a value for the still concentration X_{3s}, and with that known the partial pressures P_3 and P_4 follow from the results given in Chapter 2 and we arrive at Table 3.2.

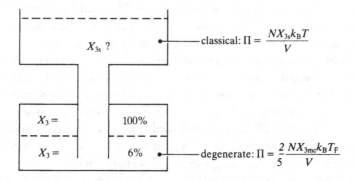

Figure 3.7. This illustrates the principle of the balancing of osmotic pressures. It is necessary to be aware, as mentioned in the text, that in the still where $X_{3s} \approx 1\%$ and $T_s \approx 0.7$ K, the dissolved ^{3}He 'gas' behaves classically, whereas in the mixing chamber where $X_{3mc} \approx 6\%$ and $T_{mc} \approx 4$ mK, it is Fermi degenerate. Appropriate expressions for the osmotic pressures have to be used.

Table 3.2 *Operating conditions in the still*

T_s (K)	X_{3s}	$(P_3 + P_4)$ (Pa)	P_3/P_4
0.50	0.014	1.9***	934
0.55	0.013	3.1**	307
0.60	0.012	4.6*	123
0.65	0.011	6.5	56
0.70	0.010	8.8	29
0.75	0.009	11.8	* 16
0.80	0.008	15.2	** 9
0.85	0.008	19.8	*** 6

* = undesirable; *** = very undesirable. Best T_s is ~ 0.7 K, where $(P_3 + P_4)$ is not too small and P_3/P_4 is not too large.

Secondly, it is important to suppress ^{4}He film flow. The figures in Table 3.2 are equilibrium values, but the mixture in the still is superfluid, and film flow can by a combination of siphoning (at a rate 0.25 μmol s^{-1} per millimetre of perimeter of pumping line) and evaporation (at a higher level where the pumping line becomes sufficiently warm) substantially increase the fraction of ^{4}He in the pumped gas. Figure 3.8 shows a 'film burning' still. Excess of ^{4}He (say, above 1%) is undesirable because

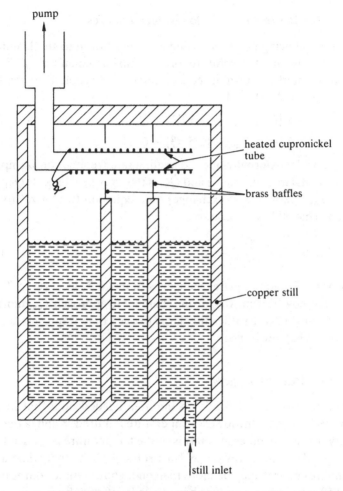

Figure 3.8. A film-inhibiting still design. The combination of the heated thin-walled cupronickel tube and the brass baffles acts to prevent the formation of film on the inside of the pumping line. The heating is not in itself a problem because the still has to be heated anyway to maintain it at a suitable temperature. (Refs: Black *et al.* (1969) and Wheatley *et al.* (1971).) A more recent design is described by Vermeulen and Frossati (1987).

(i) it loads the pumping system without producing additional cooling,
(ii) it may lead to phase separation in the concentrated return line, with the release of latent heat,
(iii) bubbles of dilute phase in the concentrated return line may behave turbulently.

3.7 How to obtain the lowest temperatures

We use the following notation: C denotes 'concentrated' and D denotes 'dilute', T_{mc} is the temperature of the mixing chamber, T_r is the temperature of the returning concentrated ^{3}He (cooled by the heat exchanger, see below), and

$$\dot{Q}_{ext} = \dot{n}_3[H_D(T_{mc}) - H_C(T_r)] - \dot{Q}_{h1} \qquad (3.4)$$

$$= \dot{n}_3[94.5 \ T_{mc}^2 - 12.5 \ T_r^2] - \dot{Q}_{h1} \qquad (3.5)$$

where $\dot{Q}_{ext}$ is the externally-applied heating rate (related for example to an experimental measurement) and $\dot{Q}_{h1}$ is the stray heat leak. If $\dot{Q}_{h1} = 0$, and $T_r = T_{mc}$ (perfect heat exchange) then equation (3.5) gives for the minimum achievable temperature,

$$T_{mc}(min) = \left(\frac{\dot{Q}_{ext}}{82\dot{n}_3}\right)^{1/2}. \qquad (3.6)$$

The purpose of a good heat exchanger is to make T_r as close to T_{mc} as possible (thereby allowing the mixing chamber to approach $T_{mc}(min)$) by the use of contact areas sufficiently large to overcome Kapitza boundary resistance to the flow of heat.

3.8 Heat exchangers

The structure of these depends on the area needed, and that in turn depends very much on the lowest temperature required. Kapitza thermal boundary resistance increases sharply as the temperature is reduced (see for example the general review by Harrison (1979)) so that while a few square metres of area may be more than enough for temperatures down to about 50 mK, it is necessary to have an order of magnitude more to get below about 10 mK. More quantitative results are given in the calculations below. Where relatively small areas are needed, exchangers in the form of coiled concentric tubes are adequate. On the other hand, where relatively large areas are needed the design must contain large contact areas in small volumes and typically includes units made of sintered silver or copper powder. It is not uncommon to have a concentric exchanger immediately below the still, followed lower down by four or more sintered units as shown in Figure 3.9.

Various ways of calculating the necessary areas can be given, but the following way offers the possibility of easy explicit computation. It makes use of a calculation model illustrated in Figure 3.10.

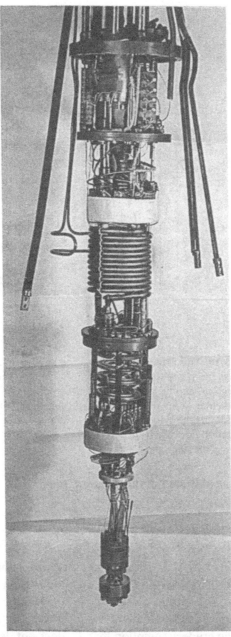

Figure 3.9. Photograph (courtesy of Mr Alan Young) of a dilution unit supplied by the Oxford Instrument Company (UK), showing (from top to bottom) the still, a coiled tubular heat exchanger, several sintered units, and the mixing chamber.

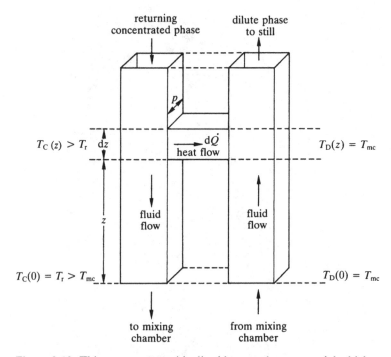

Figure 3.10. This represents an idealised heat exchanger model which is analysed in the text. Two pipes are shown with square cross-section, and heat exchange is pictured as taking place perpendicular to the fluid flows and through a slab of material. Only an element of the slab is shown at height z and with rectangular cross-section $p\,dz$. The rest of the slab is shown as transparent merely to simplify the diagram.

We assume, as indicated in the diagram, that

(i) $T_C(0) = T_r > T_{mc}$,   (3.7)

(ii) $T_C(z) > T_C(0)$,   (3.8)

(iii) $T_D(z) = T_D(0) = T_{mc}$.   (3.9)

(i) and (ii) are obvious in view of the flow direction, but (iii) is a simplifying approximation based on the observation (see Table 3.1 on p. 29) that the dilute phase has a considerably larger specific heat than the concentrated phase at the same temperature. We use this fact to justify the crude assumption that T_D does not depend on z but is constant and equal to T_{mc}, that is, that the dilute phase on the way up absorbs heat from the returning concentrated phase without showing a temperature rise. Clearly one way of improving the model is to allow T_D

as well as T_C to depend on z but we will not do that here, preferring to keep the algebraic expressions more explicit at the expense of some precision. (One must not in any case depend too heavily on model calculations where empirical experience is vitally important.) Next we make the further simplifying assumption that the simple algebraic forms for the enthalpies appropriate to full Fermi degeneracy apply not only in the mixing chamber at T_{mc} but all the way up through the heat exchanger to the still at T_s; this is not as injudicious as might at first appear since most of the area we are attempting to calculate is in fact at the lowest end of the range of temperatures in the whole device. This assumption allows us (with the help of the table of properties above) to write:

$$\text{(iv)} \quad d\dot{Q} = \dot{n}_3 dH_C \text{ with } H_C = 12.5\, T_C^2, \tag{3.10}$$

$$\text{(v)} \quad \frac{\dot{Q}}{\dot{n}_3} = 94.5\, T_{mc}^2 - 12.5\, T_r^2. \tag{3.11}$$

Finally we take as a convenient expression of the form that the Kapitza boundary conductance takes:

$$\text{(vi)} \quad \frac{d\dot{Q}}{dz} = \lambda_i p (T_C^i - T_{mc}^i), \tag{3.12}$$

where i indicates the particular power law being assumed and λ_i is independent of temperature. We shall give solutions for the three most commonly discussed cases, namely $i = 4$, 3, and 2. Substitute equation (3.10) into equation (3.12) and integrate from the mixing chamber to the still to obtain the required area:

$$A = pz = \frac{25\dot{n}_3}{2\lambda_i} \int_{T_r}^{T_s} \frac{d(T_C^2)}{(T_C^i - T_{mc}^i)}. \tag{3.13}$$

Case A ($i = 4$). If

$$\frac{d\dot{Q}}{dz} = \lambda_4 p (T_C^4 - T_{mc}^4) \tag{3.14}$$

then

$$A = \frac{25\dot{n}_3}{4\lambda_4 T_{mc}^2} \ln \left[\left(\frac{T_r^2 + T_{mc}^2}{T_r^2 - T_{mc}^2} \right) \left(\frac{T_s^2 - T_{mc}^2}{T_s^2 + T_{mc}^2} \right) \right]. \tag{3.15}$$

Case B ($i = 3$). If

$$\frac{d\dot{Q}}{dz} = \lambda_3 p (T_C^3 - T_{mc}^3) \tag{3.16}$$

then

$$A = \frac{25\dot{n}_3}{\lambda_3 T_{mc}} \left\{ \frac{1}{\sqrt{3}} \tan^{-1} \left(\frac{2\sqrt{3}T_{mc} (T_s - T_r)}{(2T_s + T_{mc}) (2T_r + T_{mc}) + 3T_{mc}^2} \right) \right.$$

$$\left. + \frac{1}{6} \ln \left[\left(\frac{T_s - T_{mc}}{T_r - T_{mc}} \right)^2 \left[\frac{T_r^2 + T_r T_{mc} + T_{mc}^2}{T_s^2 + T_s T_{mc} + T_{mc}^2} \right] \right] \right\} \quad (3.17)$$

Case C ($i = 2$). If

$$\frac{d\dot{Q}}{dz} = \lambda_2 p(T_r^2 - T_{mc}^2) \quad (3.18)$$

then

$$A = \frac{25\dot{n}_3}{2\lambda_2} \ln \left(\frac{T_s^2 - T_{mc}^2}{T_r^2 - T_{mc}^2} \right). \quad (3.19)$$

in all three cases $i = 4$, 3 and 2, the areas A from equations (3.15), (3.17) and (3.19) respectively apply in the temperature range $T_{mc} \leqslant T_r \leqslant T_s$. As could be expected, the two limiting cases for $T_r = T_{mc}$ and $T_r = T_s$ are the same for cases A, B and C, namely

$$A = \infty \text{ corresponds to } T_r = T_{mc} \quad (3.20)$$

and

$$A = 0 \text{ corresponds to } T_r = T_s. \quad (3.21)$$

The limit $A = \infty$ is approached in practice when A is large enough, that is, if $A \gg 25\dot{n}_3/2\lambda_4 T_{mc}^2$ (case A) or $A \gg 75\dot{n}_3/2\lambda_3 T_{mc}$ (case B) or $A \gg 25\dot{n}_3/2\lambda_2$ (case C), when equations (3.15), (3.17) and (3.19) all lead to $T_r = T_{mc}$ (perfect heat exchange) and hence via equation (3.11) back to equation (3.6) for $T_{mc}(\text{min})$.

The general algebraic procedure in all three cases is to eliminate T_r using the common equation (3.11) together with the appropriate form chosen from equations (3.15), (3.17) and (3.19). The limit $A = 0$ is not of physical interest but there is a 'lowish-A' régime which does not extend to $A = 0$ and which is of interest only in that the complicated formulae (3.15), (3.17) and (3.19) then reduce to more transparent forms which appear in the literature. For completeness we include them here:

Case A. If A lies in the range $25\dot{n}_3/2\lambda_4 T_s^2 \ll A \ll 25\dot{n}_3/2\lambda_4 T_{mc}^2$ then equation (3.15) reduces to an expression for T_r:

$$T_r^2 = \frac{25\dot{n}_3}{2\lambda_4 A}, \quad (3.22)$$

which, on substitution into equation (3.11) gives the result quoted by Frossati (1978):

$$T_{mc}^2 = \frac{(12.5)^2 \dot{n}_3}{94.5 \lambda_4 A} + \frac{\dot{Q}}{94.5 \dot{n}_3} . \tag{3.23}$$

Case B. If A lies in the range $75 \dot{n}_3/2\lambda_3 T_s \ll A \ll 75 \dot{n}_3/2\lambda_3 T_{mc}$ then equation (3.17) reduces to an expression for T_r:

$$T_r = \frac{75 \dot{n}_3}{2\lambda_3 A} \tag{3.24}$$

which, on substitution into equation (3.11), gives the result

$$T_{mc}^2 = \frac{12.5}{94.5} \left(\frac{75}{2}\right)^2 \left(\frac{\dot{n}_3}{\lambda_3 A}\right)^2 + \frac{\dot{Q}}{94.5 \dot{n}_3} . \tag{3.25}$$

This is very similar (but not identical) to the version given by Frossati (1978).

Case C. There is no advantage in seeking an approximation in this case since the equations have relatively simple exact forms. Equation (3.15) can be arranged to read

$$T_r^2 = T_{mc}^2 + (T_s^2 - T_{mc}^2) \exp\left(-\frac{2\lambda_2 A}{25 \dot{n}_3}\right) \tag{3.26}$$

and substitution into equation (3.19) gives the result

$$T_{mc}^2 = \frac{12.5 T_s^2 \exp(-2\lambda_2 A/25 \dot{n}_3) + \dot{Q}/\dot{n}_3}{12.5 \exp(-2\lambda_2 A/25 \dot{n}_3) + 82} . \tag{3.27}$$

Note that in this case, unlike cases A and B, T_s remains in the formulae and cannot be removed by making it arbitrarily large compared to T_r. This observation has no true physical meaning, being merely an artifact of the model.

In order to use these results it is necessary to review the evidence regarding the index i (4, 3, or 2) and the measured value of the corresponding constant λ_i. This is by no means a straightforward matter and the serious reader would be well advised to consider the detailed evidence reviewed by Frossati (1978) and Vermeulen and Frossati (1987). Relevant factors importantly include the choice of material (e.g. copper, silver, etc.), various methods of construction (e.g. coaxial tubular, sintered powder, etc.) and the temperature range itself. Moreover the model we have chosen to analyse in some detail here is most

Table 3.3 *Calculated contact areas for a range of possible mixing chamber temperatures*
(See the text for the assumed input parameters)

T_{mc} /mK	T_r /mK	A/m^2
2	4.27	178
3	7.49	100
4	10.4	71.3
5	13.3	55.6
6	16.1	45.7
8	21.7	33.6
10	27.3	26.6
12	32.8	21.9
14	38.3	18.6
16	43.9	16.1
18	49.4	14.2
20	54.9	12.7
24	65.9	10.4
28	76.9	8.73
32	87.9	7.51
36	98.9	6.55
40	110	5.79
44	121	5.17
48	132	4.65
52	143	4.21
56	154	3.83
60	165	3.50

directly applicable to tubular exchangers and strictly needs some modification for application to the commonly used groups of sintered units. For the didactic purposes of this chapter, however, we shall make use of the formulae as a way of making a sensible assessment of the design considerations. Frossati presents the evidence supporting the use of the following values in various specified circumstances.

$$\lambda_4 = 62.5 \qquad \text{W m}^{-2} \text{ deg}^{-4}, \tag{3.28}$$

or

$$\lambda_3 = 6.72 \times 10^{-3} \text{ W m}^{-2} \text{ deg}^{-3}, \tag{3.29}$$

or

$$\lambda_2 = 2.08 \times 10^{-4} \text{ W m}^{-2} \text{ deg}^{-2}. \tag{3.30}$$

Let us take as an illustration case B (λ_3 being given by equation (3.29)), with plausible input parameters $\dot{n}_3 = 2 \times 10^{-4}$ mol s^{-1} and $\dot{Q} = 30$ nW (so that from equation (3.6) $T_{mc}(\text{min}) = 1.35$ mK), and $T_s = 0.7$ K we arrive at Table 3.3 giving for each value of T_{mc} a corresponding T_r and the necessary area A.

Such tables are easily generated by a microcomputer and the results can depend quite sensitively on the input parameters. The model used is not necessarily the best or most accurate, but it does give the flavour of the argument and the numerical results are in the right range. It shows that for the lowest temperatures to be achieved it is essential to think in terms of areas of 200 m^2 or more. Vermeulen and Frossati (1987) report that for a refrigerator designed to reach below 2 mK with the high circulation rate of 10^{-2} mol s^{-1}, they used an exchanger array with a total of more than 1000 m^2. For practical reasons this area must be contained in relatively small volumes, and this is done by sintering fine powder (copper or silver) into a porous sponge (see below).

3.9   Heat leaks

The model above, and the resulting table, contain the effects of a heat current $\dot{Q}$ which is supposed to be approximately constant and made up of a residual heat leak $\dot{Q}_{h1}$ and an externally-applied heating rate $\dot{Q}_{ext}$. Now even though it is quite possible to operate the dilution refrigerator in a single-shot manner, in which case concentrated ^{3}He is not returned and there is no heating problem at the inlet to the mixing chamber, it has been argued (Wheatley *et al.* (1968) and Wheatley *et al.* (1971)) that the practical limit is then governed by viscous generation and conduction of heat in the dilute phase at the outlet. As the dilute phase leaves the mixing chamber, heat is generated dissipatively and conducted back to the mixing chamber. A consideration of the heat balance of a section of the exit tube must include three distinct contributions. Firstly heat is generated by the viscous Poiseuille flow; secondly heat is brought into the section by conduction across its ends; thirdly the above two heating rates are balanced by the rate at which the dilute mixture absorbs heat by virtue of its enthalpy change as it traverses the section. Wheatley was able to make estimates of the various parameters involved and to conclude that there exists an 'intrinsic limit':

$$T_{mc}(\text{intrinsic}) = \frac{4}{(d_D)^{1/3}} \text{ mK} \qquad (3.31)$$

where the diameter of the mixing chamber outlet (d_D) is measured in millimetres.

Frossati (1978) has also considered these matters and has given

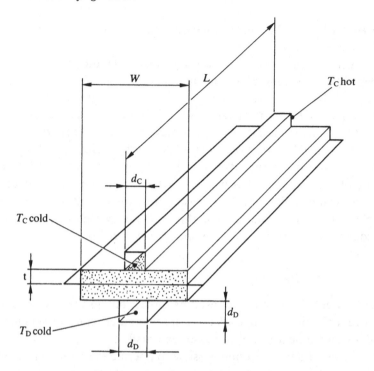

Figure 3.11. Schematic view of a heat exchanger unit made with a sintered sponge of silver powder. From Frossati (1978).

expressions including viscous and thermal conduction contributions to $\dot{Q}$ from both the dilute and concentrated sides. These expressions for $(\dot{Q}_{\mathrm{visc}} + \dot{Q}_{\mathrm{cond}})_{\mathrm{D}}$ and $(\dot{Q}_{\mathrm{visc}} + \dot{Q}_{\mathrm{cond}})_{\mathrm{C}}$ naturally depend on the geometrical characteristics of the exchangers and on the circulation rate $\dot{n}_3$. It is easy to see qualitatively that narrow channels encourage viscous heating (varying as d^{-4} for an assumed Poiseuille flow) while broad channels allow greater heat conduction from hotter regions (varying as the area πd^2). The sum of the two sources of heating thus depends on channel size as $(ad^{-4} + bd^2)$ where a and b are functions of the transport properties, temperatures, and other geometrical factors. The minimum value of $(ad^{-4} + bd^2)$ is $3(\frac{1}{4}ab^2)^{1/3}$ corresponding to an optimum channel size $d(\mathrm{opt}) = (2a/b)^{1/6}$. Frossati then makes the drastic assumption that all of this heating goes to the mixing chamber and the following Table 3.4 is worked out for a set of stepwise heat exchange units each in the form shown in Figure 3.11 each with optimised inlet and outlet diameters. The design shows a way of achieving a base temperature of 2 mK.

Table 3.4 *Calculated parameters for a set of stepwise heat exchangers designed to achieve a base temperature of 2 mK*

Stepwise exchanger:		Top	$\rightarrow$	$\rightarrow$	$\rightarrow$	Bottom
$d_C(opt)$/mm:		1.6	2.8	3.4	4.2	5.0
$d_D(opt)$/mm:		4.0	6.8	8.2	10.2	12.2
$T[^3He(C)]$/mK:	70	17.5	10.4	6.6	5.0	4.3
$T[^3He(D)]$/mK:	25	6.5	4.0	2.7	2.2	**2.0**

Frossati has shown both by calculation (as in the table above) and by experiment that it is possible, though not without the utmost attention to detail, to reach a temperature 2 mK or better. Commercial refrigerators from the Oxford Instrument Company in the UK can be obtained with base temperatures in the region of 4 mK.

3.10   Construction of heat exchangers

Coaxial tubular exchangers are often used as a first stage, but the sort of areas indicated by Table 3.3 have to be made with fine powder sintered into a porous sponge (sintering is a heat treatment carried out below the melting point, touching particles being fused at points of contact). In the simplest model of a block of sinter one thinks of a close-packed array of spherical particles for which the area per unit volume is $\pi\sqrt{2}/\delta$ where δ is the diameter of the spheres. Clearly δ must be small if areas as large as 100 m^2 are to be contained in reasonable volumes. Frossati (1978) gives experimental figures of 1.8 m^2 g^{-1} for 700 Å silver powder, sintered, and 2.2 m^2 g^{-1} for 400 Å silver powder, sintered. Figure 3.11 shows that the helium does not need to flow through the very fine pores of the sinter in order to access the large contact area. This is because the liquid has a good thermal conductivity which enables it to make use of the area as it flows through the channels of macroscopic size.

3.11   Alternative methods avoiding the need for exchangers

It is possible to bypass the problem of helium-to-solid Kapitza resistance by arranging to exchange heat as directly as possible across an interface between the hot returning concentrated ^{3}He and the cold rising dilute ^{3}He. Such methods are not widely used at present but they offer the

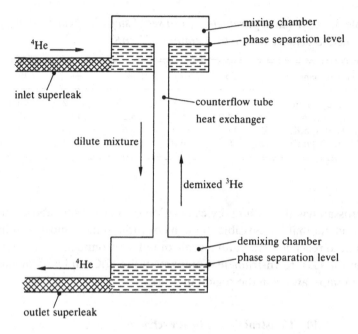

Figure 3.12. The principle of the ^{4}He-circulating Leiden refrigerator.

advantages of a simpler construction. Two alternative approaches have been put into practice. In the first (see de Waele *et al.* (1976)) ^{3}He is circulated as before but there are several mixing chambers. Temperatures of about 3 mK have been reported. In the second, known as the Leiden refrigerator, ^{4}He is circulated rather than ^{3}He. This method (see Taconis *et al.* (1971), Satoh *et al.* (1974), Jurriëns *et al.* (1977), Pennings *et al.* (1976), and Satoh *et al.* (1987)) has been successfully used as low as 3.4 mK. The principle of the Leiden refrigerator is illustrated in Figure 3.12. Superfluid ^{4}He is circulated through a superleak into the mixing chamber. The device has a counterflow tube heat exchanger in which the phase-separated descending cold mixture and the ascending hot concentrated mixture exchange heat via direct contact with each other. The counterflow tube at its lower end joins into the demixing chamber from which pure ^{4}He is taken out through an exit superleak. The heat of demixing is absorbed by a ^{3}He pot maintained at about 0.6 K. The actual arrangement used by Satoh *et al.* (1987) is shown in Figure 3.13.

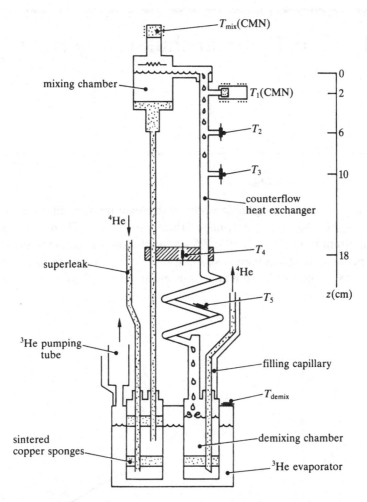

Figure 3.13. Schematic diagram of a Leiden refrigerator which can reach 3.4 mK. From Satoh *et al.* (1987).

4 The Pomeranchuk refrigerator

4.1 Preamble

Pomeranchuk cooling has few advantages over modern dilution refrigerators (see Chapter 3) or demagnetisation stages (see Chapter 5) but was important in the decade 1965–75 when dilution refrigerators and superconducting magnets were less sophisticated than they are today. It *can* cool to about 1 mK and it *is* very insensitive to applied magnetic fields.

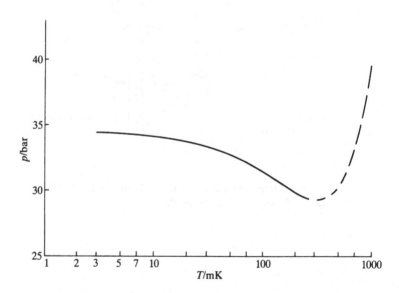

Figure 4.1. The melting curve of ^{3}He showing the unusual minimum. The broken line represents data due to Grilly (1971). Below the minimum the continuous line is derived from a polynomial fit to data by Greywall and P.A. Busch (1982); see also Greywall (1985).

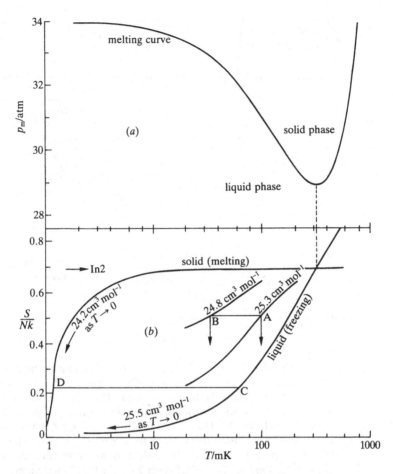

Figure 4.2. (*a*) A plot of the melting pressure, showing a minimum value of 28.93 atm at 0.32 K. (*b*) The entropies of solid and liquid ³He on the melting curve, and along two isochores.

These advantages were exploited in the historically important experiment by Osheroff *et al.* (1972) showing for the first time the low temperature phases of liquid ³He and, more recently, by for example Kopietz *et al.* (1986) and Vermeulen *et al.* (1987). On the other hand it is *not* commercially available and is *not* particularly easy to make. Also, experiments on the cooling samples, other than the refrigerant ³He itself, have not been markedly successful. For these various reasons, I shall not devote very much space to the topic.

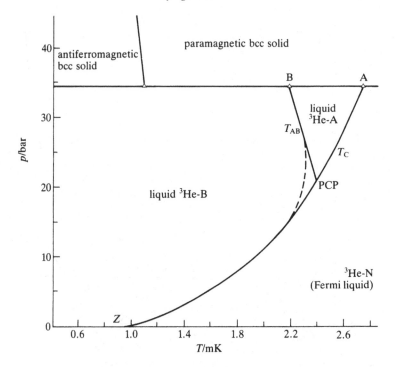

Figure 4.3. The ^{3}He phase diagram in zero applied magnetic field below about 3 mK (Halperin *et al.* (1978)). There are four labelled points: A and B mark the transitions from ^{3}He-N to ^{3}He-A and from ^{3}He-A to ^{3}He-B along the melting curve; PCP marks the polycritical point at which ^{3}He-N, ^{3}He-A and ^{3}He-B coexist; and Z marks the zero-pressure transition from ^{3}He-N to ^{3}He-B. T_{AB} denotes the transition from ^{3}He-A to ^{3}He-B, while T_c denotes the transition out of the ^{3}He-N phase. The effect upon T_{AB} of a 0.0378 tesla magnetic field is shown by the broken line.

4.2 Properties of melting helium-3

The Pomeranchuk refrigerator is based on a characteristic property of pure ^{3}He, namely, its melting curve (see Figure 4.1).

It is worth noting that the minimum in the melting curve, while not being unique, is somewhat unusual. The underlying reasons can be obtained by a consideration of the Clausius–Clapeyron thermodynamic equation which is quite general:

$$\frac{dp_m}{dT} = \frac{S_l - S_s}{V_l - V_s}. \tag{4.1}$$

The more usual situation has $V_1 > V_s$ with $S_1 > S_s$ and positive dp_m/dT. But for ^{3}He below 0.32 K, $V_1 > V_s$ as usual, but dp_m/dT is negative and we have to conclude that $S_s > S_1$. The solid is more disordered than the liquid! (The reason for this has to do with the fact that nuclear ordering occurs more readily in the liquid than in the solid.) Figures 4.2–4.4 should now be studied. It can be seen in Figure 4.2 that an isentropic compression from point A (25.3 cm^3 mol^{-1} at 100 mK) to point B (24.8 cm^3 mol^{-1}) reduces the temperature to 34 mK. Cooling by total solidification of an initially liquid sample might be represented by a horizontal line joining say, points C and D. However, as we shall see later (in particular by studying Figure 4.8), total solidification is neither necessary nor desirable for the achievement of the lowest temperatures.

4.3 Cooling by solidification

Figure 4.2(*b*) shows that cooling can be achieved by the isentropic solidification of liquid ^{3}He by compression. But there are practicalities to consider (see Figure 4.5)

The seemingly obvious method illustrated on the left hand side of Figure 4.5 does not work. The reason is that between the room temperature ^{3}He gas compressor and the low-temperature cell (provided that is below 0.32 K – see Figure 4.2(*a*)) there is bound to be a region where the temperatures are around 0.32 K and in that region the pressure will be sufficient to cause solidification in an awkward place. A block will form, and further increases of pressure at room temperature will not be transmitted. The more elaborate arrangement of bellows shown on the right hand side of Figure 4.5 is one way of overcoming this problem, and uses ^{4}He to move a low-temperature piston into the Pomeranchuk cell. In such designs and others, in which liquid ^{4}He is used hydraulically to transmit pressure to the ^{3}He through bellows or diaphragms, it is important to remember that ^{4}He has a lower freezing pressure (25 bar) than does ^{3}He (29–34 bar depending on temperature). It follows that the design must incorporate a means of maintaining a pressure differential of several bars (no less than four) between ^{4}He and ^{3}He. This may make use of geometry (e.g. of the bellows area) or elastic stiffness (e.g. of a diaphragm) or a combination.

The cooling power is given by

$$\dot{Q} = \dot{n}_3 T(S_{s,m} - S_{l,m}) \qquad (4.2)$$

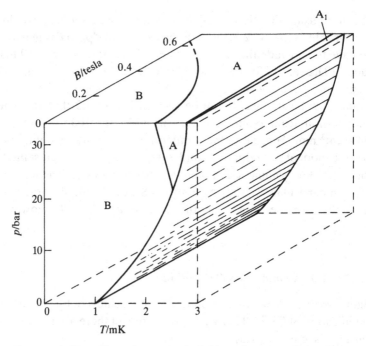

Figure 4.4. The effect of a magnetic field on the phase diagram.

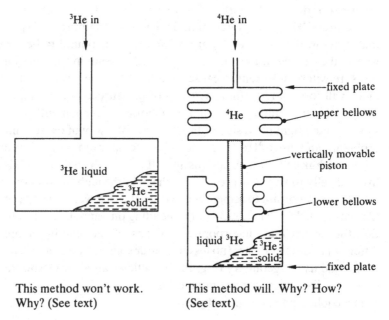

This method won't work.
Why? (See text)

This method will. Why? How?
(See text)

Figure 4.5. Practicalities.

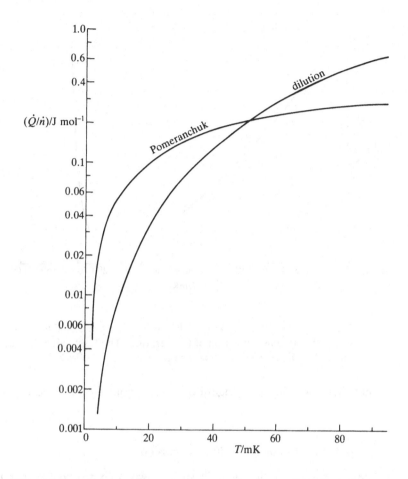

Figure 4.6. A comparison of the relative cooling powers $\dot{Q}/\dot{n}$ of the dilution refrigerator and the Pomeranchuk refrigerator. Although the techniques for inducing flow of ^{3}He across the phase boundary in the two cases are quite different, it seems clear that above about 50 mK dilution refrigeration is more efficient (mol for mol) than Pomeranchuk refrigeration but loses out at lower temperatures.

where $\dot{n}_3$ is the rate of conversion of liquid into solid. The quantity $T(S_{s,m} - S_{l,m})$ is of course simply a latent heat of freezing. The cooling power is formally analogous to that for the dilution refrigerator, $\dot{Q} = \dot{n}_3 T(S_D - S_C)$. If we were to take $\dot{n}_3$ to be the same in both, although the techniques for producing $\dot{n}$ are quite different in the two cases, we would conclude that the Pomeranchuk refrigerator is more

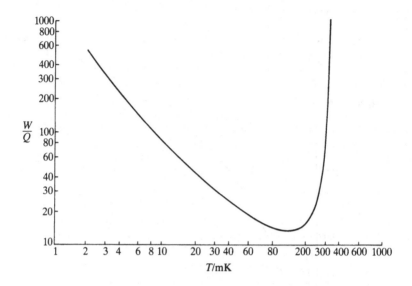

Figure 4.7. The quantity W/Q is the ratio of compressional work to heat extraction as a function of temperature. The least value is 13 at about 0.14 K, rising sharply as T approaches zero or 0.32 K.

efficient than the dilution refrigerator below about 50 mK. See Figure 4.6.

4.4   Need to be gentle in compression

The device has to be designed in such a way that the 'cold' that is generated is not cancelled or hidden by heat produced by friction in compression. Now the heat extraction produced by freezing n_3 mol of ³He is

$$dQ = n_3 T(S_{s,m} - S_{l,m}) \qquad (4.3)$$

and the work done is

$$dW = n_3 p_m(V_{l,m} - V_{s,m}), \qquad (4.4)$$

so the ratio of work done to heat extracted is found by dividing equation (4.4) by equation (4.3) and then using the Clausius–Clapeyron equation (4.2) to obtain

$$\frac{dW}{dQ} = -\frac{p_m \, dT}{T_m \, dp_m} \qquad (4.5)$$

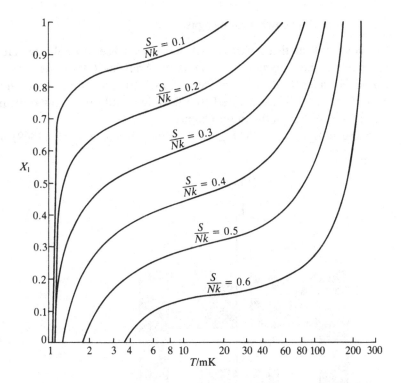

Figure 4.8. The importance of beginning compression at a low initial temperature. The lower the initial temperature, the lower is the initial entropy for a given initial liquid fraction. For $X_{1,i} = 100\%$ at 22 mK, the appropriate isentrope is $S/Nk = 0.1$ which leads down to, say 2 mK for $X_{1,f}$ as high as 82%. This high final liquid fraction avoids heating by crushing crystals.

$$= - \frac{d \ln T}{d \ln p_m} . \tag{4.6}$$

Figure 4.7 shows the magnitude of this quantity as a function of temperature.

Work itself would not matter if the work could be done reversibly. But frictional heating can be caused by rubbing, by exceeding the elastic limit of the cell containing the ^{3}He during compression, or by the crushing of ^{3}He crystals in the cell. Fortunately it is not necessary to solidify more than a fraction of the sample (see Figure 4.8) and several different geometries have been successfully used, allowing gentle compression of ^{3}He by ^{4}He. See Figures 4.9–4.11.

4.5 Some practical designs

We now look at three ways which have been used to solidify ^{3}He in practice. In the original experiments by Anufriev (1965) the ^{3}He was contained in a cell like a drum (see Figure 4.9). Basically, as ^{4}He in the central chamber is compressed, the flexible membranes bow outwards and compress ^{3}He in the outer chamber.

In the design shown in Figure 4.10 (see Johnson *et al.* (1969) and

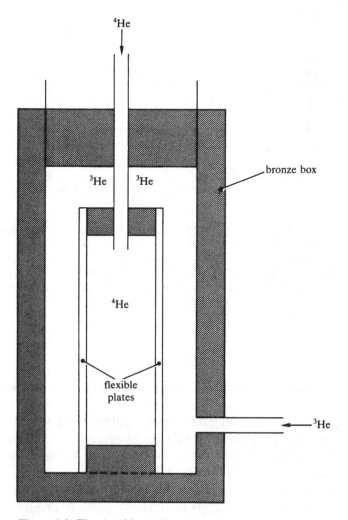

Figure 4.9. The Anufriev cell.

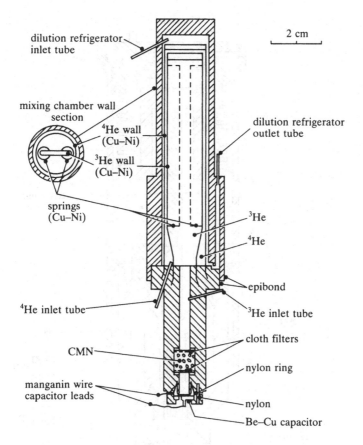

Figure 4.10. The la Jolla cell.

Johnson and Wheatley (1970)) the Pomeranchuk cell was embedded within a more elaborate construction used to precool the ^{3}He. Its chief innovation, apart from the spring-loaded ^{3}He chamber, was the fact that the cell was made integral with the mixing chamber of a dilution refrigerator. The time to precool the ^{3}He to the starting temperature was quite long, and after solidification the Pomeranchuk cell was insulated from the mixing chamber of the dilution refrigerator by the Kapitza boundary resistance. Pressure was applied through liquid ^{4}He (which does not have the pronounced minimum in the melting curve which is shown by ^{3}He). This pressure was used to squeeze a suitably flexible flattened tube, shown at the side in cross-section, and so compress the ^{3}He.

The technique most widely used after its invention at Cornell Univer-

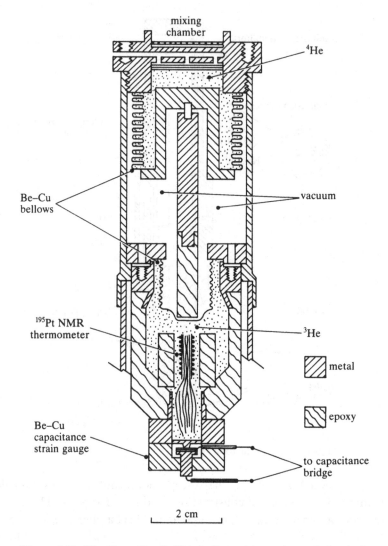

Figure 4.11. The Cornell cell. This bellows-driven piston design is due to Osheroff *et al.* (1972).

sity in the USA made use of flexible bellows as shown in Figure 4.11. As in the la Jolla cell (Figure 4.10) ^{4}He is used to transmit pressure from room temperature to the ^{3}He. The pressure from the ^{4}He expands the upper bellows which forces a contraction of the lower bellows linked by a rigid piston. As in Figure 4.5 the two bellows areas are different so as to allow for the fact that ^{4}He has a lower freezing pressure than ^{3}He does.

4.6   Some more recent designs

Kummer *et al.* (1975) used a compression cell of cylindrical shape in which the ^{3}He was held inside a hollow brass tube 2.44 cm in diameter, 10.80 cm long, with a wall thickness of 150 μm. Increasing the ^{4}He

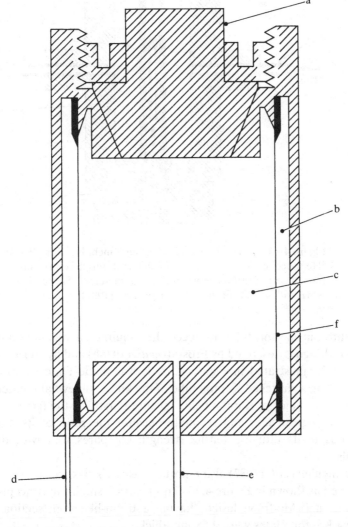

Figure 4.12. A cylindrical plastic Pomeranchuk cell: a, conical epoxy plug; b, ^{4}He space; c, ^{3}He space; d, ^{4}He capillary; e, ^{3}He capillary; f, Kapton tube. From Vermeulen *et al.* (1987).

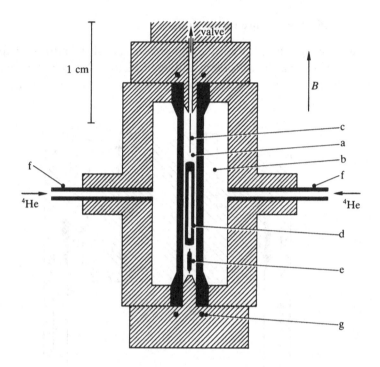

Figure 4.13. A disk-shaped plastic Pomeranchuk cell: a, ^{3}He space; b, ^{4}He space; c, vibrating wire viscometer; d, sapphire pressure transducer; e, carbon resistance thermometer; f, ^{4}He capillaries; g, NMR saddle coil. From Vermeulen *et al.* (1987).

pressure outside this tube produced the required change in volume. A similar design is described by Frossati *et al.* (1978*b*) and by Vermeulen *et al.* (1987), illustrated in Figure 4.12. It was made almost entirely of plastic with a stretched Kapton tube of thickness somewhat greater than 40 μm, the choice of Kapton being favoured firstly by the fact the NMR coils can be outside, and secondly because the ratio of its Young's modulus to its ultimate tensile strength compares well with that of metals.

Vermeulen *et al.* (1987) also report the use of a disk-shaped Pomeranchuk cell as shown in Figure 4.13. This is rather similar in conception to the original Anufriev design, having a drum-like construction. The flexing Kapton plates were 0.55 mm thick.

The use of Kapton has also been reported by Kopietz *et al.* (1986) who used the cell shown in Figure 4.14 in experiments on spin polarisation of ^{3}He in which there was necessarily a constant field of 8 tesla present. The

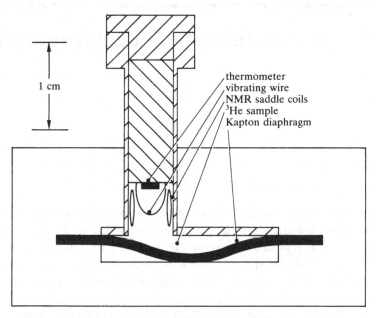

Figure 4.14. This shows the design by Kopietz *et al.* (1986), in which
the volume of the ^{3}He sample was altered by a change in the pressure
of liquid ^{4}He occupying the volume below the Kapton diaphragm.

idea was again to apply pressure from liquid ^{4}He to ^{3}He through a flexible
(Kapton) diaphragm. The vibrating wire and the NMR coils are for
purposes related to the experiment in hand.

4.7 Conclusions

Such great progress has been made in the techniques of refrigeration by
nuclear demagnetisation (see Chapter 5) that it is doubtful if the
Pomeranchuk refrigerator will ever again be widely used as once seemed
possible. One of the major problems is that where specimens *other than*
^{3}He at its melting pressure are required to be cooled, the cooling must be
done *indirectly*, that is, through a thermal link. In a compression, it is
known (Halperin *et al.* (1974), and Johnson *et al.* (1970)) that solid will
form preferentially on hot or rough areas or areas where solid has already
formed. So as heat from the cooling specimen passes to the cell through
the link it will encourage a build-up of solid in just the place where heat
flow needs to be good. But solid ^{3}He is a poor thermal conductor and so
the heat flow is blocked. This problem has yet to be satisfactorily solved;
meanwhile progress has been made elsewhere.

5   Adiabatic nuclear demagnetisation

5.1   Preamble

The modern dilution refrigerator often has a base temperature within the low range 2–4 mK, depending on design (see Chapter 3). It has thus to a large extent replaced the technique originally introduced in the early 1930s which involved the adiabatic demagnetisation of electronic paramagnetic salts such as for example chromic potassium alum ($Cr_2(SO_4) \cdot K_2SO_4 \cdot 24H_2O$) and cerous magnesium nitrate ($2\ Ce(NO_3)_3 \cdot 3\ Mg(NO_3)_2 \cdot H_2O$). The latter (CMN), and its relative CLMN in which some of the magnetic cerium ions are replaced by non-magnetic lanthanum, can reach temperatures below 2 mK and consequently have been common choices until quite recently. A further improvement, demonstrated by Šafrata et al. (1980), involves the substitution of deuterium for hydrogen in the water of crystallisation (see Table 5.1).

Table 5.1 *X denotes the fraction of cerium ions replaced by lanthanum. The initial conditions in all cases were $B_i = 1.85$ tesla and $T_i = 0.95$ K. The final magnetic temperatures in all cases were uncertain to within ± 0.07 mK*

	Final magnetic temperature (mK)	
X	CLMN	Deuterised CLMN
1.00	1.70	1.47
0.22	1.16	0.98
0.10	0.87	0.67
0.05	0.67	0.42

From Šafrata et al. (1980).

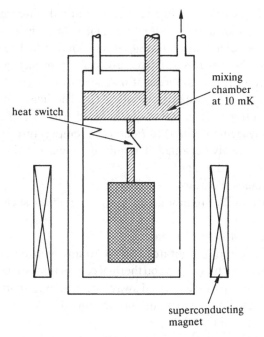

Figure 5.1. Diagram illustrating the full cycle of magnetisation and demagnetisation.

Table 5.1 shows clearly that the temperatures obtainable with CMN, CLMN, and deuterised CLMN extend well below the base temperatures of almost all dilution refrigerators. In competition with nuclear paramagnets, however, salts have been largely overtaken in practice for the purposes of adiabatic demagnetisation. The reasons for this have to do mainly with cooling power and a range of design considerations. In order to constrain the length of this book, it has therefore been decided to omit this topic, instead referring the reader to other reviews mostly written before nuclear demagnetisation had spread far beyond its originating laboratories. See Hudson (1972), Lounasmaa (1974), Betts (1976), and/or White (1979).

5.2 Basic ideas

The state of the art of nuclear demagnetisation in 1982 is reviewed in some detail by Andres and Lounasmaa (1982) and the reader who wishes to go beyond the contents of this chapter should turn to that article.

The two most important nuclear refrigerants are (i) copper, and (ii) $PrNi_5$ (an intermetallic compound of praesodymium and nickel; $PrNi_5$ is pronounced 'praesodymium nickel five').

The basic steps are illustrated in Figure 5.1. The initial state may be taken as $T_i = 10$ mK (achievable by contact through a closed heat switch with the mixing chamber of a dilution refrigerator) with $B = 0$. The steps are as follows. (i) The field is raised from zero to B_i (typically using a superconducting solenoid), and the heat of magnetisation is conducted through the heat switch to the mixing chamber. (ii) The heat switch is opened so that the refrigerant is thermally isolated. (iii) The field is reduced slowly (ideally isentropically) to B_f and the temperature falls to T_f. The initial state is easily regained if required by closing the heat switch and reducing the field to zero.

The procedure includes the following items:
— use copper or PrNi$_5$ mounted as a construction such as a 'bundle',
— magnetise with $B_i \approx 10$ tesla,
— cool with dilution refrigerator to $T_i \approx 10$ mK with superconductive heat switch closed (good thermal conduction) so that the heat of magnetisation is carried away by the mixing chamber,
— insulate 'bundle' by opening superconductive heat switch (poor thermal conduction),
— reduce B_i to $B_f \approx 0.25$ tesla over a period $\tau \approx 8$ h,
— find $T_f < 1$ mK.

5.3 Entropy data

First we present in Figures 5.2–5.4 graphical information about the dependence of the entropy on both temperature and applied magnetic

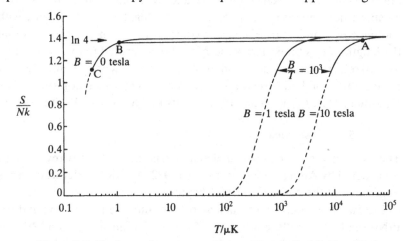

Figure 5.2. Nuclear spin entropy of copper for applied fields of 0, 1, and 10 tesla (from Betts, 1976).

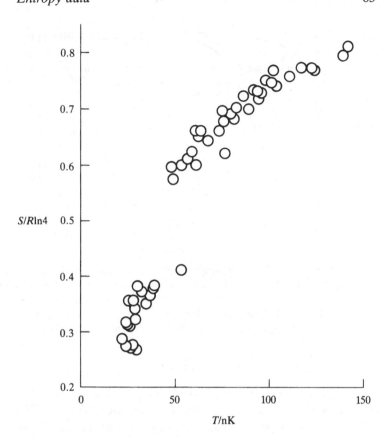

Figure 5.3. Entropy of copper below 150 nK.

field for both copper and PrNi$_5$. In Figure 5.2 this nuclear spin entropy of copper for applied fields of 0, 1 and 10 tesla is shown. An isentropic demagnetisation from point A ($B_i = 10$ tesla, $T_i = 35$ mK, $B/T_i \approx 300$ T/K) produces a final temperature of order 1 µK at point B. Significantly higher values of B_i/T_i are not very practical, and B_i/T_i ≈ 2000 (e.g. 10 tesla at 5 mK) is probably close to a practical upper limit. It follows that in practice in single-stage demagnetisations only small reductions in entropy are achieved, so that the restricted form of the equation given below is almost always adequate. The entropies for $B = 1$ and 10 tesla are represented by broken curves.

In Figure 5.3 the entropy of copper is shown at the lowest temperature so far reported. Taken from Huiku *et al.* (1986), these data were obtained in a two-stage demagnetisation cryostat (see Ehnholm *et al.* (1980)). In Figure 5.4 the entropies of copper and PrNi$_5$ are compared.

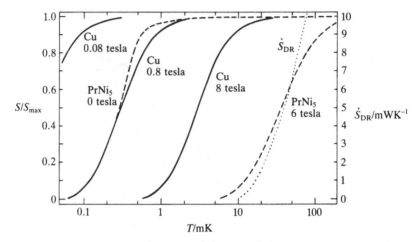

Figure 5.4. Nuclear entropy of copper (full lines) and PrNi$_5$ (broken lines) as a function of temperature for various external fields, compared to the rate of entropy reduction $\dot{S}_{DR}$ of a typical dilution refrigerator (dotted line). For copper, $S_{max} = 11.5$ J deg^{-1} mol^{-1} and for PrNi$_5$, $S_{max} = 14.9$ J deg^{-1} mol^{-1}. From Pobell (1982).

5.4 Ideal nuclear paramagnet

It is helpful to have some relatively simple algebraic expressions as reasonable approximations, and the following are often used. They apply rather well for copper in most situations relevant for present purposes. For PrNi$_5$ they can be used, but only subject to the observations made in Section 5.7 'hyperfine-enhanced Van Vleck paramagnets'.

We suppose that we can treat our magnetic refrigerant as a solid assembly of N weakly interacting nuclear magnetic moments at temperatures above the ordering temperature. The entropy S_n of such an assembly at sufficiently high temperatures and low fields approaches the familiar form $Nk_B \ln(2z + 1)$ corresponding to isotropic disorder of the moments, where z is the appropriate spin quantum number. A better form, allowing for finite temperature and field dependences, is

$$\frac{S_n}{Nk} = \ln(2z + 1) - \frac{C}{2k_B \mu_0} \frac{V}{N} \frac{(B^2 + B_{int}^2)}{T^2} \qquad (5.1)$$

where C is the Curie constant in $\chi = C/T$, and B_{int} is the internal field, which can for present purposes be regarded as a small adjustable parameter. For example if as a method of obtaining B_{int} copper we simply take from Huiku *et al.* (1986) the value $S_n = 0.65 \, Nk \ln 4$ at

$T = 60$ nK we deduce that $B_{int} = 7.2 \times 10^{-5}$ tesla. This is somewhat smaller than the widely-quoted earlier figure of 3.1×10^{-4} tesla but for most practical purposes the difference is not of great significance. With this substitution, the numerical form of equation (5.1) appropriate to copper is

$$\frac{S_n}{Nk} = \ln 4 - \frac{1.76 \times 10^{-15}}{T^2} - 3.4 \times 10^{-7}\left(\frac{B}{T}\right)^2 \qquad (5.2)$$

and this covers almost all practical situations. An adiabatic (isentropic) demagnetisation gives from equation (5.1)

$$T_f = T_i \left(\frac{B_f^2 + B_{int}^2}{B_i^2 + B_{int}^2}\right)^{1/2}. \qquad (5.3)$$

5.5 Reality (non-ideality)

The above expressions assume that there are no heat leaks or dissipative processes associated with demagnetisation. At least three such effects exist in reality:

  (i) heat leak,
  (ii) eddy current heat production because nuclear refrigerants are, as metals, good conductors of electricity,
  (iii) spin–lattice relaxation.

If (iii) were absent, the effects of (i) and (ii) could be minimised by a suitable choice of demagnetisation time Δt. For example Gylling (1971) shows that for a linear demagnetisation from B_i to B_f which causes during the interval Δt an eddy current heating rate $q\dot{B}^2$ (where q is a constant containing resistivity and geometry), the following relation holds:

$$\frac{B_i}{T_i} - \frac{B_f}{T_f} = \frac{\mu_0}{CV}\left(\frac{\dot{Q}\Delta t}{(B_i - B_f)} + \frac{q(B_i - B_f)}{\Delta t}\right)\ln\left(\frac{B_i}{B_f}\right) \qquad (5.4)$$

where $\dot{Q}$ is a constant heat leak and B_{int} is taken as zero to simplify. It readily follows that there is a best demagnetisation interval Δt_{best} which minimises $[(B_i/T_i) - (B_f/T_f)]$. This is

$$\Delta t_{best} = (B_i - B_f)(q/\dot{Q}) \qquad (5.5)$$

for which

$$\left(\frac{B_i}{T_i} - \frac{B_f}{T_f}\right)_{min} = \frac{\mu_0(q\dot{Q})^{1/2}}{CV}\ln\left(\frac{B_i}{B_f}\right). \qquad (5.6)$$

Experimentally one attempts to restrict $\dot{Q}$ to the level of a nanowatt and to keep q small by suitable geometrical design which might for example make use of wires of powder geometry. It is generally not desirable to prepare the refrigerant deliberately so as to reduce its electrical conductivity (and hence q) because that would also reduce the thermal conductivity and thus impair the ability of the refrigerant to cool a specimen or cell. This is because of the Wiedemann–Franz law which links the electrical and thermal conductivities. Indeed copper refrigerant constructions are often heat-treated so as to improve the resistance ratio R_{res} (defined as the ratio of electrical resistivity at room temperature to the residual resistivity at the lowest temperatures); the thermal conductivity is then given approximately by $K = 1.5 \, R_{res} T$ W m^{-1} deg^{-1}. For more details see for example Betts *et al.* (1978). But item (iii) is also very important and it happens that a too-rapid demagnetisation may be undesirable not only because of the effects of eddy current heating as mentioned above but also because it can cause the temperature T_n of the nuclei to separate from the temperature T_e of the electrons-plus-lattice. This may not matter if one is interested in the properties of those nuclei themselves. But one cannot solder samples directly to nuclei, only to lattices!

5.6   Spin–lattice relaxation

The simplest model (ignoring eddy current heating) is represented in Figure 5.5. Given the details (i.e. $\dot{Q}$ and $B(t)$) of the influences on T_e and T_n (see Figure 5.5 and its rubric), two coupled equations can be written down for $T_e(t)$ and $T_n(t)$. These both include a term describing the Korringa mechanism (Korringa (1950)) which is expressible as

$$\frac{d(1/T_n)}{dt} = - \frac{(1/T_n) - (1/T_e)}{\tau_1} \tag{5.7}$$

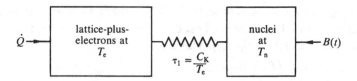

Figure 5.5. Heating and cooling of the lattice-plus-electrons. The lattice-plus-electrons at T_e gain heat from heat leaks and exchange heat with the nuclei at T_n via the Korringa mechanism. The temperature T_n of the nuclei is sensitive to the applied field $B(t)$ and the nuclei also exchange heat with the electrons-plus-lattice via the Korringa mechanism.

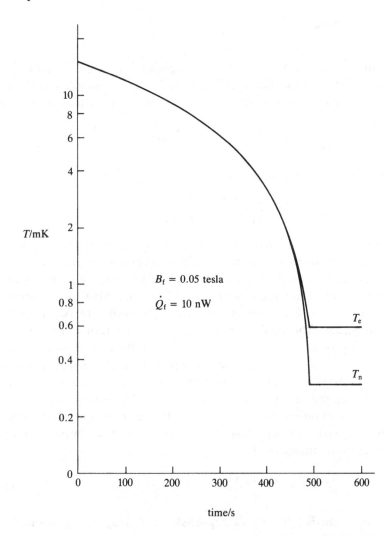

Figure 5.6. Calculated solution of equations (5.10) and (5.11) for T_e and T_n during and immediately after a linear demagnetisation from 16 mK and 0.6 to 0.05 tesla in 490 s of a 10 cm³ sample of copper assuming a heat leak of 10 nW. From Hensel (1973).

where

$$\tau_1 T_e = C_K \tag{5.8}$$

where C_K is the Korringa constant. For copper $C_K = 0.4$ s K in zero field and $C_K = 1.1$ s K in fields greater than about 0.1 millitesla. Rearrangement of equations (5.7) and (5.8) leads to

$$\frac{dT_n}{dt} = \frac{T_n(T_e - T_n)}{C_K}. \tag{5.9}$$

This result, together with the heat capacity of the nuclei (deduced from equation (5.1) for their entropy, with $B_{int} = 0$ for simplicity) allows the two coupled equations to be written down in a fairly transparent form:

$$\frac{\dot{Q}}{Nk} = \gamma T_e \frac{dT_e}{dt} + \frac{C}{k\mu_0 C_K} \frac{V}{N} B^2 \frac{(T_e - T_n)}{T_n} \tag{5.10}$$

for the electrons-plus-lattice, and

$$\frac{dT_n}{dt} = \frac{T_n(T_e - T_n)}{C_K} + \frac{T_n}{B} \frac{dB}{dt} \tag{5.11}$$

for the nuclei. The solution is not explicit and must be done numerically; Figure 5.6 shows an example of such a calculation by Hensel.

The initial field B_i is usually determined by a variety of considerations importantly including cost and the need to control or eliminate the fringing field at the experimental cell to be cooled by contact with the refrigerant. The final field B_f is usually finite rather than zero. The reasons for this are that if it is too low then the nuclear heat capacity is small (see equation 5.1) for the entropy so that after demagnetisation T_n is rapidly raised even by small heat leaks; if on the other hand it is too high then the nuclei are never effectively cooled. A simplistic calculation suggests an optimum final field $B_f(\text{opt})$ which results in a minimum final temperature $T_e(\text{min})$. Equation (5.10) with a further simplification $\gamma = 0$ gives on rearrangement

$$\frac{T_e}{T_n} = 1 + \frac{\mu_0 C_K}{CB_f^2} \frac{\dot{Q}}{V}. \tag{5.12}$$

Now, immediately after demagnetisation we may use equation (5.3) to approximate

$$T_n = T_i B_f / B_i \tag{5.13}$$

and hence reduce equation (5.12) to the form

$$T_e \left(\frac{B_i}{T_i} \right) = B_f + \frac{\mu_0 C_K}{CB_f} \frac{\dot{Q}}{V} \tag{5.14}$$

and this is readily minimised with respect to B_f to give

$$B_f(\text{opt}) = \left(\frac{\mu_0 C_K}{C} \frac{\dot{Q}}{V} \right)^{1/2} \tag{5.15}$$

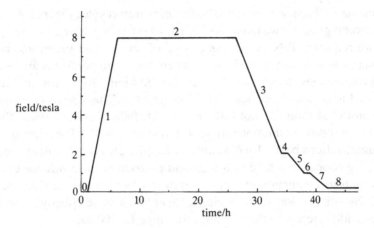

Figure 5.7. A magnetisation/demagnetisation cycle used at Sussex. The numerals 0–8 merely label the various stages of the cycle. The field in the final stage 8 was not zero, for reasons discussed above, but 0.25 tesla.

together with

$$T_e(\text{min}) = 2 \left(\frac{T_i}{B_i}\right) B_f(\text{opt}) = 2T_n(\text{min}). \tag{5.16}$$

As $\dot{Q}$ causes T_e and T_n to rise from their minimum values $T_e(\text{min})$ and $T_n(\text{min})$, the relationship is governed by equation (5.12) and we see that

$$T_e(t) = 2T_n(t). \tag{5.17}$$

But these equations are all approximations and users learn more in practice by mathematical modelling (e.g. Hensel (1973), Betts *et al.* (1978), Dow *et al.* (1982), Gachechiladze *et al.* (1986)) and direct experience. Figure 5.7 shows a typical magnetisation and demagnetisation schedule as used at one stage of the work at Sussex. The schedule was conveniently automated by use of a microcomputer.

5.7 Hyperfine-enhanced Van Vleck paramagnets

The usefulness of these materials, importantly including $PrNi_5$, was first indicated by Al'tshuler (1966). What happens in simple terms when an external field is applied to $PrNi_5$ is that the electronic configuration is adjusted in such a way as to produce a greatly magnified field at the nucleus; the effect is called hyperfine enhancement. In the case of $PrNi_5$ the external field is enhanced by a factor of about 14, so the entropy curves for PrNi$_5$ are shifted to higher temperatures (see Figure 5.4). This is advantageous in two main ways. First, where temperatures of order 1 mK are required, PrNi$_5$ gives the chance of smaller and therefore cheaper magnets and cold ends. Second, where lower temperatures are needed, PrNi$_5$ bridges an awkward gap between dilution refrigerators and copper used as a second stage, for the following reason. The copper stage typically consists of about 20 mol with an available field of about 8 tesla. But the cooling power of dilution refrigerators (see Chapter 3) decreases as T^2 and usually drops below 1 µW at around 10 mK. This is unfortunate because using copper in a field of 8 tesla and precooling to 15 mK for example results in a reduction of only 4% of the nuclear entropy (see Figure 5.2). PrNi$_5$ offers a way of precooling copper much more effectively and thus providing access to the temperature range 10–100 µK.

5.8   Geometry of the refrigerant: plates, wires, or powder?

In the early days of nuclear demagnetisation using copper, it was believed that fine-grained geometry was essential because it inhibited eddy current heating. The wire bundle has been widely used in consequence, although the wire diameter has tended to grow from early extreme fineness to the substantial proportions of more recent rod diameters of order 2 mm or more. The main reason for this is constructional convenience where demagnetisation times of several hours are in any case determined by other considerations of reversibility in cooling samples. If one considers a rod of radius r and electrical resistivity ϱ in a magnetic field directed along its axis and changing at a constant rate $\dot{B}$, one can show that the heating generated per unit volume by eddy currents is

$$\frac{\dot{Q}}{V} = \frac{\dot{B}^2 r^2}{8\varrho} = \frac{R_{res}\dot{B}^2 r^2}{8\varrho_{RT}} \tag{5.18}$$

where ϱ_{RT} is the resistivity at room temperature and R_{res} is the resistance ratio mentioned above (in Section 5.5.6: *reality (non-ideality)*) and which depends on the purity and mechanical history of the sample. Using some reasonable numerical values ($\dot{B} = 0.25$ tesla h^{-1} as in the final stages of the schedule illustrated in Figure 5.7, $r = 1$ mm) and applying them to carefully prepared copper ($R_{res} = 600$, $\varrho_{RT} = 1.55 \times 10^{-8}$), equation (5.18) divided by the density of copper gives $\dot{Q}/M = 2.6$ nW kg^{-1}. This heat load ceases after the demagnetisation ends and is acceptable for most purposes.

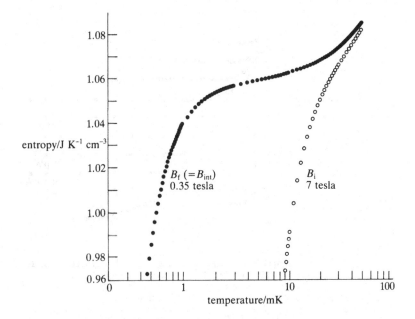

Figure 5.8. The entropy of 35% pure ^{3}He (at a pressure of 33 bar) and 65% copper flakes in a field of 7 tesla and in a final internal field of the flakes of 0.35 tesla. A temperature of 0.85 mK can be achieved by a demagnetisation from 15 mK, and 1.5 mK can be reached from a 20 mK starting temperature. From Bradley *et al.* (1984).

Thus wires and rods of copper have been used, and so have blocks machined into a finned or otherwise convoluted geometry. PrNi$_5$ is not available in such a variety of forms as copper, and has been most frequently used in the form of rods. The question of eddy current heating is of much less importance since the electrical conductivity is much lower than that of copper. The rods of PrNi$_5$ are sometimes coated with cadmium which acts as a non-superconducting solder above 3 millitesla.

When the refrigerant is to be used to cool other materials to the region of 1 mK or lower, notably liquid ^{3}He and metallic samples, problems of thermal boundary resistance arise. The Lancaster group (Bradley *et al.* (1984)) have developed new methods for cooling into the microkelvin régime. This method, applied to the important case of ^{3}He, involves the immersion of powdered copper refrigerant in the liquid. The powder was in the form of flakes (thickness ≈ 1 μm, width ≈ 25 μm, designated Bronze–Copper Super 1000 by the manufacturer Wolstenholme Bronze Powders, Manchester, England). The entropy of a mixture of pure ^{3}He

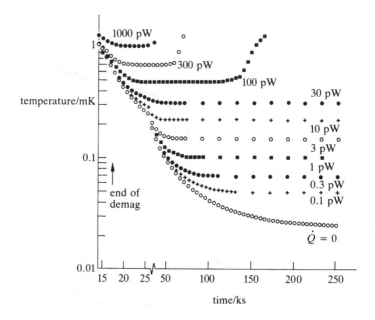

Figure 5.9. The time evolution after demagnetisation of the temperature of a 6.4% ³He/⁴He solution mixed with 20% by volume of copper powder, as a function of the total heat leak into the liquid. From Bradley *et al.* (1984); see also Dow *et al.* (1982).

and these flakes was calculated and is shown in Figure 5.8 which shows the potential effects.

In practice much depends, as mentioned above, on the various boundary resistances between ³He and copper nuclei, and the group carried out some computer simulations, one of which is shown in Figure 5.9

Bradley *et al.* (1984) suggest that for some purposes a very simple cell design would suffice, as shown in Figure 5.10; they themselves used a rather more elaborate design as shown in Figure 5.16. They described the construction of one version of the double cell used in the latter design as follows:

the refrigerant in the inner cell is in the form of nine 99.99 + % pure copper plates, each 1 mm × 18 mm × 50 mm. The copper is silver plated and each flat face carries an approximately 0.5 mm layer of sintered silver powder. The sinter is made by adding approximately equal weights of nominally 700 Å powder and nominally 0.3 μm powder compressed to filling factor of about 50% and sintered for 10 min at 200 °C in a hydrogen-rich atmosphere. The total contact area in the inner cell is estimated to be 13 m². The outer cell is filled with the

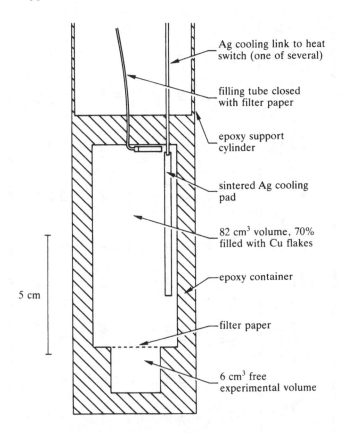

Figure 5.10. A simple experimental cell for use with copper flakes. The ratio of copper to helium is the same as that assumed in the calculated entropy of Figure 5.8. From Bradley *et al.* (1984).

unannealed copper flakes. To increase the copper content, the flakes are precompressed and small lumps of the pressed material tamped into the outer space to give a filling factor of around 30%.

5.9 Apparatuses

A very useful review of the methods and apparatuses used up to 1982 has been given by Andres and Lounasmaa (1982) and Table 5.2 is reproduced from that source. Some examples of successful nuclear demagnetisation cryostats are shown in Figures 5.11–5.16.

Table 5.2 *Examples of copper nuclear refrigerators for cooling liquid ^{3}He (up to 1981). From Andres and Lounasmaa (1982), who collected much of the information by private communication*

	Bell Labs (Osheroff and Sprenger 1980)	Berkeley (Eisenstein et al. 1979)	Cornell (Archie 1978)	Grenoble (Frossati 1978)	La Jolla (Krusius et al. 1978)	Los Angeles (Bozler et al. 1978)	North-western (Mast et al. 1980)	Ohio State (Muething et al. 1979)	Orsay (Avenel et al. 1976)	Otaniemi (Veuro 1978)	Sussex (Hutchins 1981)
Nuclear stage											
n (mol)	15	9	21	8	17.5	14	19	40	45	25	15
Diam. of wires (mm)	0.2	0.23	0.5	0.6	0.25	0.20	0.34	0.18	0.5	0.2	0.5
T_i (mK)	16	19	19	7	14	16	20	27	12	18	20
B_i (rms value, T)	6.7	8	7	8	5.5	8	6.4	7.7	8	5.3	8
B_f (rms value, mT)	20	50	28	100	1.5	50	50	—	30	30	80
T_3 (min)(mK)	0.22	0.48	0.34	0.4	0.41	0.35	0.39	0.4	0.31	0.38	0.7
$\dot{Q}_c$ (nW)	0.4	1.5	<1	100	1.5	1	1.5	5	3	0.9	—
$\dot{Q}_3$ (nW)	1	2	—	0.5	—	—	—	0.3	1	0.2	3.5
$R_K T_3^2$(K^2/W)	7	50	—	21	—	40	—	60	50	90	—
$R_c T_c^2$ (K^2/W)	7	0.4	—	4	—	30	—	5	—	30	—
Warm-up to 2 mK (h)	—	150	>100	30	>100	>100	100	—	>100	100	—
^{3}He cell											
Amount of ^{3}He (cm^3)	18	4	15	25	8.9	11	17	16.7	9	17	10
Type of sinter	Ag	Ag	Cu	Ag	Cu	Cu	Cu	Pd	Cu	Ag	Cu
Particle size (nm)	100	70	30	40	flakes	flakes	1000	2000	2000	100	1000
Surface area (m^2)	100	40	107	60	215	45	35	25	30	10	8

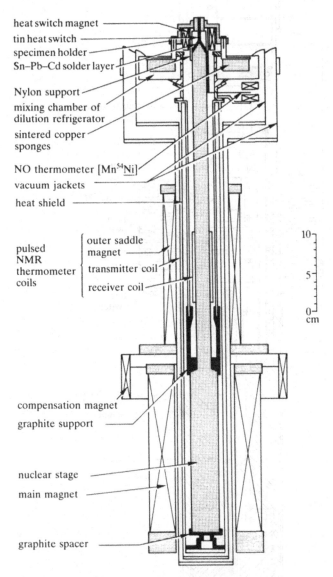

Figure 5.11. An apparatus for the single-stage nuclear
demagnetisation of copper. From Berglund, Ehnholm *et al.* (1972).
12 mol of copper were used, a lowest electron temperature of
0.37 mK was achieved and it was possible to keep the nuclear stage
below 2 mK for 12 h.

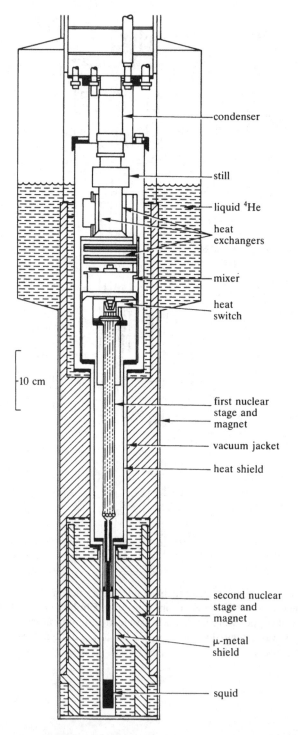

Figure 5.12. An apparatus for the two-stage nuclear demagnetisation of copper. From Ehnholm *et al.* (1980).

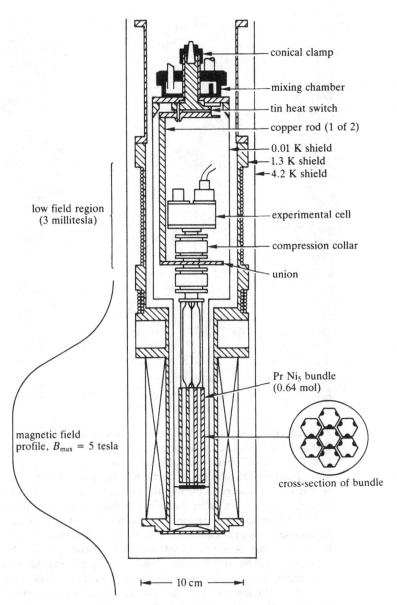

Figure 5.13. An apparatus for the single-stage nuclear demagnetisation of PrNi$_5$. From Greywall (1985). The refrigerant consisted of 0.64 mol of PrNi$_5$ in the form of seven 8 mm diameter hexagonal rods. It was able to cool ^{3}He samples to less than 0.3 mK.

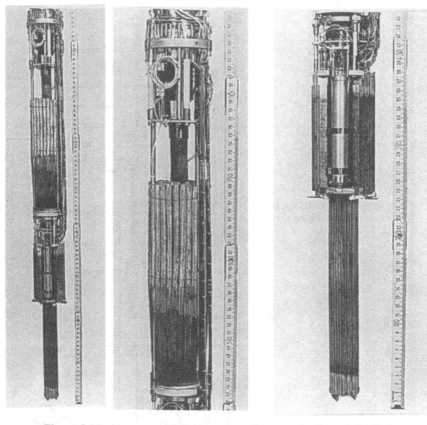

Figure 5.14. An apparatus for two-stage demagnetisation in which the first stage is PrNi$_5$ and the second is copper. From Mueller *et al.* (1980). See also Pobell (1982). (*a*) Low temperature part of the apparatus with the dilution refrigerator at the top, the PrNi$_5$ nuclear demagnetisation stage in the middle, and the copper stage at the bottom of the picture immediately below the experimental region. (*b*) PrNi$_5$ nuclear demagnetisation stage, with the coil for the upper superconducting heat switch just below the bottom plate of the mixing chamber. A bundle of 500 copper wires runs through the centre of the PrNi$_5$ stage, connecting its bottom plate to the upper heat switch. Six copper wires were soldered with cadmium to each PrNi$_5$ rod. The lower ends of these wires were welded into the bottom plate. (*c*) Copper nuclear demagnetisation stage, welded to the bottom of the experimental region. The two nuclear stages were precooled to about 25 mK by a dilution refrigerator. The first stage was 4.29 mol (1.86 kg) of PrNi$_5$ in an initial field of 6 tesla. This was sufficient to cool the copper stage to about 5 mK in a field of about 8 tesla. After demagnetisation, electron temperatures of 48 μK could be achieved and experiments performed below 60 μK for a few days.

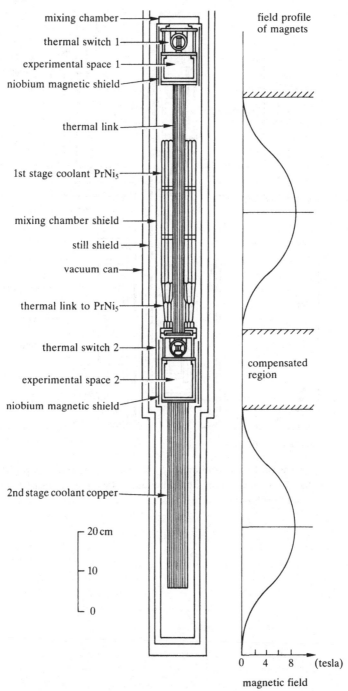

Figure 5.15. Another apparatus for two-stage nuclear demagnetisation in which the first stage is PrNi$_5$ and the second, which is reported to have reached 27 μK, was copper. From Ishimoto *et al.* (1984).

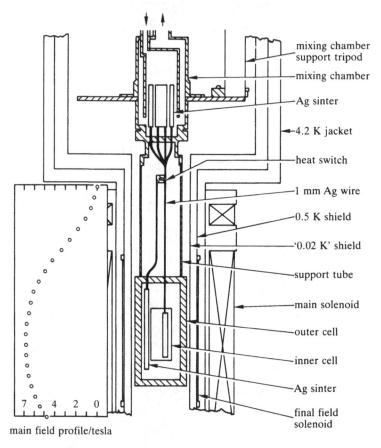

main field profile/tesla

Figure 5.16. From Bradley *et al.* (1984). This design uses powdered copper immersed in the liquid helium to be cooled, as discussed in the text on pages 72–3.

6 Thermometry

6.1 Preamble

The theoretical bases of thermometry are discussed in Chapter 1. It is rare for experimentalists to use methods which relate directly to a Carnot cycle, though there are certainly users of secondary methods which have respectable theoretical support of another sort (e.g. the Boltzmann factor in ^{60}Co γ-ray anisotropy). Some properties of some materials have useful dependences on temperature in the desired range (e.g. the electrical resistance of carbon composites) but are best described by polynomial fits rather than fundamental theory. The fact is that there are many ways of measuring temperatures in the millikelvin range and experimentalists make choices which best fit their needs. The extents to which these techniques give temperatures which are accurate (i.e. equal to the true Kelvin temperatures) or consistent when compared with temperatures determined in other ways are often not fully certain. Users frequently rest their cases on widely accepted results obtained and published by respected workers in the past (e.g. for the melting pressure of ^{3}He (Greywall (1985)), or they take strength from demonstrations by others that two methods do in fact agree within acceptable limits. Examples of comparisons include platinum NMR versus nuclear orientation (Berglund, Collan et al. (1972)), platinum NMR versus CLMN susceptibility (Alvesalo et al. (1980)), noise versus ^{60}Co γ-ray anisotropy (Soulen and Marshak (1980)), and ^{3}He melting pressure versus CLMN susceptibility (Parpia et al. (1985)). Disputes can and do arise about whether small temperature-dependent effects are truly properties of the system being studied or whether they are artifacts of an imperfect scale. A recent example of this has been the debate about whether the low temperature heat capacity of liquid ^{3}He does or does not contain a term proportional to $T^3 \ln T$. These are important matters to be clear about and progress is being made, but for the majority whose interests are not

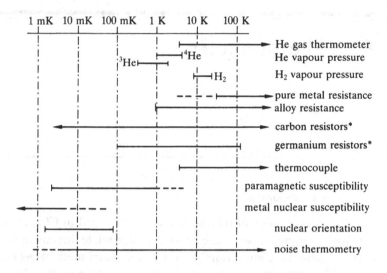

Figure 6.1. Temperature ranges of various thermometer types. The asterisk indicates where a single device does not cover the whole range. From McClintock *et al.* (1984).

primarily in the specialist areas of thermometry the best advice is to choose the most appropriate practical methods, to read the research literature thoroughly on that method, and to write up results in a form which gives all relevant details for future reference and, if necessary, correction.

Good general references include Lounasmaa (1974), Hudson *et al.* (1975), and Betts (1976), although it is noticeable that all of these were written more than a decade ago. In each of the sections of this chapter the reader will find at least one more recent reference providing an entry to the more recent literature. Figure 6.1 shows the temperature ranges covered by some of the methods available.

We shall discuss the following methods:
(i)   NBS superconducting fixed point device (Section 6.2)
(ii) Helium vapour pressure (Section 6.3)
(iii) ^{3}He melting pressure (Section 6.4)
(iv) Carbon and germanium resistance (Section 6.5)
(v)   Capacitance (Section 6.6)
(vi) CMN and CLMN magnetic susceptibility (Section 6.7)
(vii) Platinum nuclear magnetic susceptibility and spin-lattice relaxation time (Section 6.8)
(viii) ^{60}Co γ-ray anisotropy (Section 6.9)
(ix) Resistor noise (Section 6.10)

The discussions will be fairly brief, but key references are intended to direct the reader to more detailed expositions if these are required.

6.2 The NBS superconducting fixed point device

Table 6.1 *Summary of properties of SRM 768*

Superconducting material	Transition temperature will lie between mK	Typical transition width mK	Typical reproducibility[a] upon thermal cycling mK
W	15.0–17.0	0.7	0.20
Be	21.0–24.0	0.2	0.10
Ir	99.5–99.5	0.8	0.10
AuAl$_2$	160.0–161.0	0.3	0.10
AuIn$_2$	205.0–208.0	0.4	0.15

[a] The values given are the averages of the standard deviations of several samples.

This device (code-named the SRM 768) is intended by the US National Bureau of Standards for use as a means of calibrating other thermometers (see Soulen and Dove (1979)). It consists of a few lengths of different superconducting metals whose transition temperatures in zero field are referable to NBS standards. The measurement techniques used by NBS for this purpose are mainly γ-ray anisotropy (see Section 6.9 below) and noise thermometry (see Section 6.10 below). Users should keep in touch with the NBS in case further work necessitates the provision of a modified table. The transitions are however listed in Table 6.1 for reference.

The device is provided with two sets of coils, each set consisting of a primary and secondary coil. The eight leads from these sets are interconnected in series opposition, however, and thus require only four leads for measurement. The superconductive transitions are observed as changes in the mutual inductance of the coils which occur when the magnetic field is expelled from the interior of the sample as it enters the superconducting state (the Meissner effect). The mutual inductance change may be observed using for example a Hartshorn bridge. There are some fairly stringent recommended operating conditions which should be adhered to in using the device if the best precision is to be achieved. These are as follows:

(i)   peak-to-peak   magnetic   field   applied   in   primary   coil:
      2.3 microtesla for the tungsten (W) transition, 0.46 microtesla
      for the others,
(ii) heating generated with above conditions: 1.8 nW and 75 pW
      respectively,
(iii) ambient magnetic field kept below 1 microtesla.

6.3   Vapour pressure of helium-3

As mentioned in Chapter 1, the current (but still provisional) scales for
^{3}He and ^{4}He are jointly titled EPT-76. Durieux and Rusby (1983) and
Rusby (1985) give for the vapour pressure of ^{3}He in the range 0.2 K to T_c
(3.3 K),

$$\ln p = b \ln T + \sum_{k=-1}^{4} a_k T^k \qquad (p \text{ in Pa}, T \text{ in K}) \qquad (6.1)$$

where $b = 2.254\ 84$, $a_{-1} = -2.509\ 43$, $a_0 = 9.708\ 76$, $a_1 = -0.304\ 433$,
$a_2 = 0.210\ 429$, $a_3 = -0.054\ 5145$, and $a_4 = 0.005\ 6067$.

At 0.2 K (the lower limit of applicability), $p = 0.001\ 473$ Pa, a
pressure which can only be measured with extreme care. An inversion of
this formula has been given by Rusby and Durieux (1984): For ^{3}He in the
range $T = 0.2$–2.0 K,

$$T = \sum_{i=0}^{8} b_i [(\ln(p) - A)/B]^i \qquad (p \text{ in Pa}, T \text{ in K}) \qquad (6.2)$$

where $A = 1.8$, $B = 8.2$, $b_0 = 0.426\ 055$, $b_1 = 0.432\ 581$, $b_2 = 0.380\ 500$,
$b_3 = 0.299\ 547$, $b_4 = 0.213\ 673$, $b_5 = 0.149\ 533$, $b_6 = 0.099\ 716$,
$b_7 = 0.044\ 546$, and $b_8 = 0.007\ 914$.

Measurement can be done by a combination of mercury and oil
manometers and a McLeod gauge; the ultimate accuracy of such a
method is probably (see Carr (1964)) about 1% at 10^{-3} Torr (0.1% in T
at $T = 0.28$ K), 3% at 10^{-4} Torr (0.2% in T at $T = 0.23$ K), and 7% at
10^{-5} Torr (0.4% in T at $T = 0.20$ K). The equipment needed is elaborate
and breakable, and also suffers from the disadvantage that the pressure is
measured at room temperature and so must be corrected for thermo-
nuclear effects. It is preferable in most circumstances to use a diaphragm
whose flexure under pressure from one side is detected capacitatively; a
suitable cell is shown in Figure 6.2.

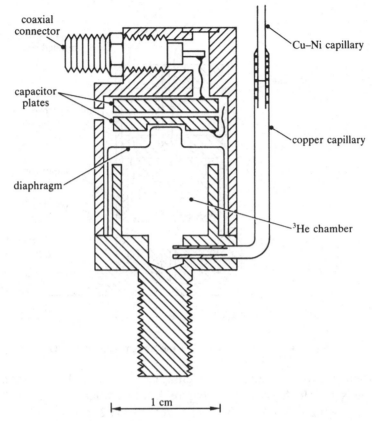

Figure 6.2. The vapour pressure cell described by Greywall and Busch (1980). The pressure resolution is about 10^{-5} Torr, and at $T = 0.3$ K and $T = 2.0$ K this results in temperature resolutions of 0.3 mK and 0.04 μK respectively.

6.4 Helium-3 melting pressure

The dependence of melting pressure on temperature is shown in Figure 4.1. The pressure is best measured by a cell like the vapour pressure cell shown above, and a suitable design by Greywall (1985) is shown in Figure 6.3. The movement of the diaphragm is measured capacitatively and its response is calibrated with a dead weight tester. With a pressure resolution of 10 μbar, the precision of temperature measurements is 3 in 10^4 at 1 mK, 3 in 10^5 at 10 mK, 5 in 10^6 at 100 mK.

For ^{3}He in the range 1–250 mK, Greywall gives the following formula:

$$p = p_A + \sum_{i=-3}^{5} a_i T^i \qquad (p \text{ in bar}, T \text{ in mK}) \qquad (6.3)$$

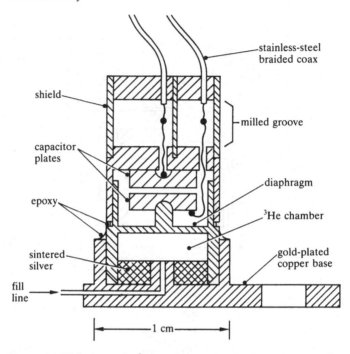

Figure 6.3. This shows the ^{3}He melting pressure cell of Greywall (1985).

where p_A is the pressure on the melting curve at which the transition from ^{3}He-N to ^{3}He-A occurs (see Figure 4.3) and where $p_A = 34.3380$ bar, $a_{-3} = -0.260\ 784\ 92 \times 10^{-1}$, $a_{-2} = 0.843\ 248\ 81 \times 10^{-1}$, $a_{-1} = -0.109\ 908\ 60 \times 10^0$, $a_0 = 0.151\ 204\ 00 \times 10^0$, $a_1 = -0.450\ 703\ 32 \times 10^{-1}$, $a_2 = 0.173\ 702\ 24 \times 10^{-3}$, $a_3 = -0.521\ 411\ 83 \times 10^{-6}$, $a_4 = 0.125\ 616\ 45 \times 10^{-8}$,   and $a_5 = -0.140\ 515\ 00 \times 10^{-11}$.

If the measurement has a pressure resolution of 10 μbar, the precision of temperature measurements is 3 in 10^4 at 1 mK, 3 in 10^5 at 10 mK, 5 in 10^6 at 100 mK. If you cannot reach as low as T_A you may prefer an alternative earlier form by Greywall (1983) which refers to the minimum p_{min} in the melting curve rather than p_A. In this earlier version, which applies in the range 3 mK to 330 mK, the formula suggested is

$$p = p_{min} + \sum_{i=-2}^{5} a_i T^i \qquad (p \text{ in bar}, T \text{ in mK}) \qquad (6.4)$$

where $p_{min} = 29.3175$ bar, $a_{-2} = 2.1895 \times 10^{-8}$, $a_{-1} = -8.1989 \times 10^{-5}$, $a_0 = 5.162\ 54$, $a_1 = -44.0395$, $a_2 = 153.846$, $a_3 = -350.634$, $a_4 = 594.115$, and $a_5 = -465.947$.

6.5   Carbon or germanium resistance

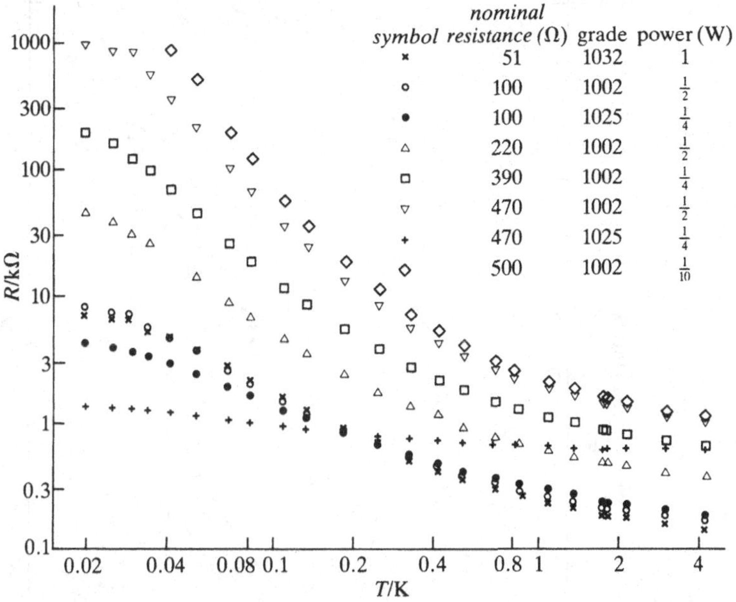

Figure 6.4. This shows the resistances of a selection of Speer resistors (see Black *et al.* (1964)). The saturation at the lowest temperatures is almost certainly due to self-heating.

This method is easy to use but is not a wise choice for absolute thermometry. It was reviewed (up to 1975) by Betts (1976). Many experimenters still use early Allen–Bradley resistors originally described by Clement and Quinnell (1952) for which one may use the approximate formula

$$T \approx \frac{1.8786}{\log_{10} R + 3.4576/\log_{10} R - 3.7243} \quad (T \text{ in K, } R \text{ in } \Omega). \quad (6.5)$$

The constants vary from sample to sample, of course, and it is usual to calibrate to find better values of the three numerical constants. At the lowest temperatures Allen–Bradley resistors become too high, whereas Speer resistors have more suitable characteristics; their performances are as shown in the Figure 6.4.

More recently, Matsushita resistors have increasingly found favour (see Kobayasi *et al.* (1976) and Koike *et al.* (1985)). A variety of fitting formulae have appeared in the literature; a useful recent paper is by Sullivan and Edwards (1986).

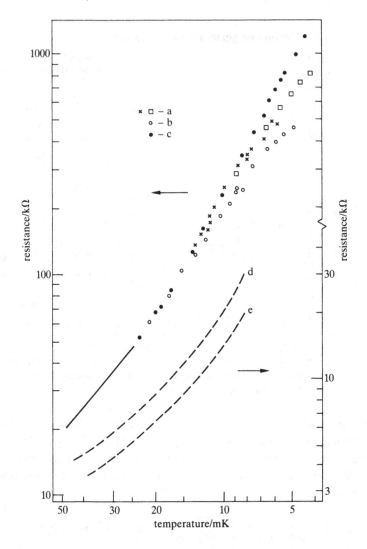

Figure 6.5. This shows resistance versus temperature results by Eska and Neumaier (1983). The different runs represented were as follows. (*a*) Speer resistor, T1-NMR thermometer; (*b*) and (*c*) Speer resistor, Cu-susceptibility thermometer; (*d*) shielded Matsushita resistor; (*e*) unshielded Matsushita resistor. The contributions of the leads are not subtracted.

With all carbon resistors, saturation due to self-heating ultimately limits the sensitivity to temperature but with careful preparation and very low power (4 fW) resistors have been used to 4 mK by Eska and Neumaier (1983) as shown in Figure 6.5. The importance of limiting the measuring power becomes much more marked as the temperature falls, and for many purposes powers of 1 nW or greater are perfectly acceptable. It is in principle a simple matter of deciding what degree of temperature sensitivity is required for the experiment in hand and then designing or buying a bridge which can deliver that performance. Since the saturation is related to thermal boundary resistances there are some grounds for using an estimate that the upper limit on the power varies as T^3. Taking the figure mentioned above (4 fW at 4 mK) would lead to a rule that $P \leqslant 60\ T^3$ nW but it would be unwise to use this as anything more than a very rough guide in advance of an experimental test.

The carbon resistors mentioned in this section are not manufactured specially for low temperature thermometry, and they are usually cheap unless there has been a change in the manufacturing process which reduces the desirability of the new ones and increases the rarity value and hence the purchase price of the older ones. Germanium resistors, on the other hand, are elaborately made specifically for thermometry. They are more expensive, but since they can be bought with referable calibrations they are quite commonly used, though not at the lowest temperatures.

6.6 Capacitance

Some dielectric materials show changes of capacitance with temperature, and this is an attractive proposition in view of the precision with which capacitance can be measured. The record has been a little discouraging however until a recent paper by Reijntjes *et al.* (1986) who investigated the low-frequency capacitance of a glass thermometer made of very pure SiO_2 doped with approximately 1200 ppm OH^-. This was investigated in the range 15–40 mK in magnetic fields up to 6 tesla. Within their temperature accuracy (using ^{60}Co γ-ray anisotropy – see Section 6.9 – of ±5% no shift of the capacitance due to the applied fields was observed. This insensitivity to magnetic field is of course an advantage. The temperature sensitivity is shown in Figure 6.6.

6.7 Cerous magnesium nitrate (CMN and CLMN)

We also refer to CLMN, in which some of the magnetic cerium atoms in CMN are replaced by non-magnetic lanthanum. The procedure is to take a powdered sample and measure its magnetic susceptibility. Then

$$\frac{1}{T - \Delta} = A(\chi - \chi_0). \tag{6.6}$$

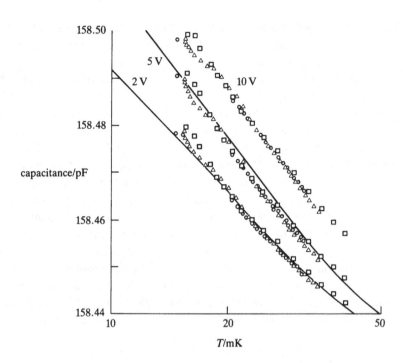

Figure 6.6. The temperature dependence of a glass thermometer at three different measuring voltages but all at the same frequency 4.7 kHz. The circles, triangles, and squares were taken at 0.0, 0.25 and 6.0 tesla respectively. From Reijntjes *et al.* (1986).

Calibration against well known temperatures (e.g. the ^{3}He vapour pressure (see Section 6.3 above) or the NBS superconducting fixed point device (see Section 6.2 above) allows a determination of A, χ_0 and Δ. Below about 8 mK, Δ is not accurately constant and perhaps rises to about 0.3 mK at $T = 2$ mK in CMN. CLMN is better in this respect, though its susceptibility is smaller because of the reduced number density of magnetic cerium ions. Parpia *et al.* (1985) have used CLMN with a ratio of 4–5% cerium to lanthanum with the expectation that the ordering temperature would be suppressed well below 1 mK while maintaining a reasonable Curie constant. Figure 6.7 shows the dependence of the susceptibility of CLMN on the temperature derived from a ^{3}He melting curve thermometer.

Traditionally, χ was measured by mutual inductance methods. But Greywall and Busch (1982) used a self-inductance bridge at 10 kHz and more recently a new edition of the 'Robinson oscillator' has been

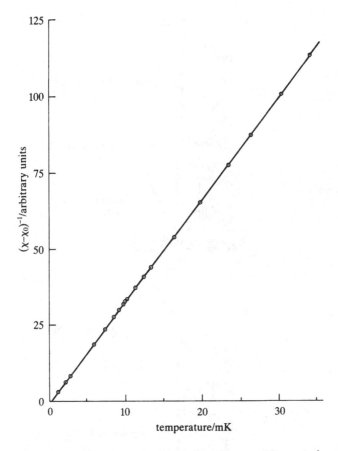

Figure 6.7. This shows the linear dependence of $(\chi - \chi_0)^{-1}$ on temperature derived from a ^{3}He melting curve thermometer. From Parpia *et al.* (1985).

reported (Robinson (1987)). Cell designs are shown in Figure 6.8. Recent developments in squid magnetometry can also be applied.

6.8   NMR methods

Some metals, notably platinum and copper, have nuclear magnetic susceptibilities. They are small and need the extra sensitivity of NMR methods but since platinum (for example) is well described as an ideal paramagnet to below 1 mK, it obeys Curie's law $\chi = C/T$ rather accurately. Moreover a completely independent property, the Korringa relation $\tau_1 T = C_K$, can also be used to double-check the results although

CMN thermometer

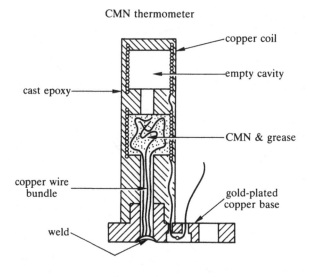

copper coil

empty cavity

cast epoxy

CMN & grease

copper wire bundle

gold-plated copper base

weld

CLMN thermometer

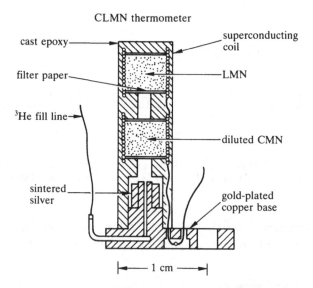

cast epoxy

superconducting coil

filter paper

LMN

^{3}He fill line

diluted CMN

sintered silver

gold-plated copper base

|—— 1 cm ——|

Figure 6.8. This shows the CMN and CLMN thermometers used by Greywall and Busch (1982).

early optimism that the two methods would give exactly the same temperature has been dampened by more recent experience. The geometry of an NMR head is shown in Figure 6.9.

For examples of the use of platinum NMR thermometry see Aalto *et*

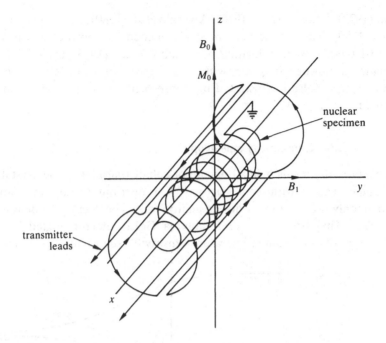

Figure 6.9. The geometry of an NMR head.

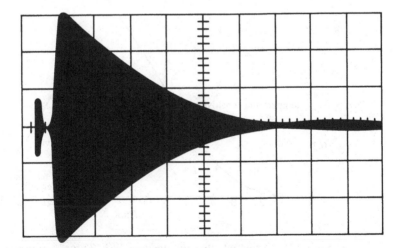

Figure 6.10. A typical free induction decay signal was from a platinum thermometer measured at 25 mK. The time scale was 200 μs per division. From Aalto *et al.* (1973).

al. (1973), Ahonen *et al.* (1976), Alvesalo *et al.* (1980), and Osheroff and Yu (1980). The sample must be finely divided into powder or wires to restrict eddy current heating and Aalto *et al.* (1973) used 3.7 g of platinum powder 0–10 μm with a frequency of 413 kHz. A decay following a pulse is illustrated in Figure 6.10 and can be analysed to obtain $(\chi - \chi_0) \propto 1/T$.

6.9   Gamma-ray anisotropy

This typically needs a small needle of cobalt (parallel to the crystallographic *c*-axis) with some active ^{60}Co distributed (an alternative commonly-used source is ^{54}Mn in an iron foil). The decay scheme is well known. The detector is (preferably) a GeLi detector angled in the direction of maximum temperature sensitivity (see Figure 6.11). There

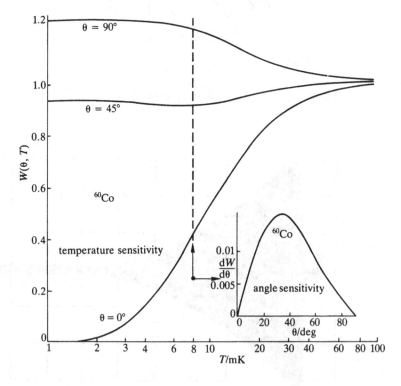

Figure 6.11. Evaluated intensity functions $W(\theta, T)$ for ^{60}Co in a cobalt host in zero applied field for $\theta = 0°$, 45° and 90°, assuming point detectors. The insert shows the angle sensitivity. For details of the formulae see for example Lounasmaa (1974) or Betts (1976).

are two γ-ray lines and one can use either or both. It is easy (but not particularly cheap) to use a multichannel analyser linked to a micro-computer for direct read-out of temperature. *But*, good data come from lengthy counting times, which can be a disadvantage. Good references are Berglund *et al.* (1972), and more recently, Marshak (1983, 1986).

6.10   Noise thermometry

The idea is simple: the mean square Johnson noise voltage appearing, within a bandwidth $\Delta\omega$ of angular frequencies for which $h\omega \ll k_B T$, across a complex impedance $Z = \mathrm{Re}\, Z(\omega) + \mathrm{j}\, \mathrm{Im}\, Z(\omega)$ is

$$\Delta\langle V_n^2 \rangle = \frac{2k_B T}{\pi} \,\mathrm{Re}\, Z(\omega)\, \Delta\omega. \tag{6.7}$$

This is as respectable theoretically as the perfect gas law, so if you can measure $\Delta\langle V_n^2 \rangle$ you have access to an absolute thermometer. Unfortunately the instrumental sophistication required has inhibited wider use, partly because $\Delta\langle V_n^2 \rangle$ is small (e.g. with $\Delta\omega = 2\pi \times 10^6$ Hz, $T = 1$ K and $\mathrm{Re}\, Z = 1$ MΩ, the RMS voltage is 7 μV), and partly because other types of noise must be eliminated with care. Good references are by Webb *et al.* (1973), Soulen and Giffard (1978) and Seppä (1986).

6.11   Conclusion

Although each section in this chapter is quite brief, the reader can easily make further progress by turning to the general references or to the specific references given in the sections.

References

Aalto, M.I., Collan, H.K., Gylling, R.G. and Nores, K.O. (1973). *Rev. Sci. Instrum.*, **44**, 1075.

Abel, W.R., Johnson, R.T., Wheatley, J.C. and Zimmerman Jr, W. (1967). *Phys. Rev. Lett.*, **18**, 737.

Ahonen, A.I., Krusius, M. and Paalanen, M.A. (1976). *J. Low Temp. Phys.*, **25**, 421.

Al'tshuler, S.A. (1966). *Zh. Eksp. i Teor. Fiz. Pis'ma* **3**, 177; English translation in *Sov. Phys. J.E.T.P. Lett.*, **3**, 112 (1966).

Alvesalo, T.A., Berglund, P.M., Islander, S.T., Pickett, G.R. and Zimmerman Jr, W. (1971). *Phys. Rev.*, **A4**, 2354.

Alvesalo, T.A., Haavasoja, T., Manninen, M.T. and Soine, A.T. (1980). *Phys. Rev. Lett.*, **44**, 1076.

Anderson, A.C., Edwards, D.O., Roach, W.R., Sarwinski, R.E. and Wheatley, J.C. (1966). *Phys. Rev. Lett.*, **17**, 367.

Andres, K. and Lounasmaa, O.V. (1982). Chapter 4 (entitled 'Recent progress in nuclear cooling', *Progress in Low Temperature Physics*, volume 8, ed. D.F. Brewer (Amsterdam: North-Holland) p. 221.

Anufriev, Yu.D. (1965). *Sov. Phys. J.E.T.P. Lett.*, **1**, 155.

Berglund, P.M., Collan, H.K., Ehnholm, G.J., Gylling, R.G. and Lounasmaa, O.V. (1972). *J. Low Temp. Phys.* **6**, 357.

Berglund, P.M., Ehnholm, G.J., Gylling, R.G., Lounasmaa, O.V. and Søvik, R.P. (1972). *Cryogenics*, **12**, 297.

Betts, D.S. (1976). *Refrigeration and thermometry below one kelvin* (Brighton: Sussex University Press).

Betts, D.S., Brewer, D.F., Cox., A.J., Hutchins, J., Saunders, J. and Truscott, W.S. (1978). Proceedings of the International Institute of Refrigeration Conference (Zürich, Switzerland, 1978) entitled 'Advances in refrigeration at the lowest temperatures', p. 163.

Black, Jr, W.C., Hirschkoff, E.C., Mota, A.C. and Wheatley, J.C. (1969). *Rev. Sci. Instrum.*, **40**, 846.

Black, Jr, W.C., Roach, W.R., and Wheatley, J.C. (1964). *Rev. Sci. Instrum.*, **35**, 587.

Bradley, D.I., Bradshaw, T.W., Guénault, A.M., Keith, V., Locke-Scobie, B.G., Miller, I.E., Pickett, G.R. and Pratt Jr, W.P. (1982). *Cryogenics*, **22**, 296.

Bradley, D.I., Guénault, A.M., Keith, V., Kennedy, C.J., Miller, I.E., Musset, S.G., Pickett, G.R. and Pratt Jr, W.P. (1984). *J. Low Temp. Phys.*, **57**, 359.
Bureau International des Poids et Mesures. (1979). The 1976 provisional 0.5 K to 30 K temperature scale, *Metrologia*, **15**, 65.
Carr, P.H., (1964). *Vacuum*, **14**, 37.
Clement, J.R. and Quinnell, E.H. (1952). *Rev. Sci. Instrum.*, **23**, 213.
Comité International des Poids et Mesures. (1969). The international practical temperature scale of 1968, *Metrologia*, **5**, 35.
de Bruyn Ouboter, R., Taconis, K.W., le Pair, C. and Beenakker, J.J.M. (1960). *Physica*, **26**, 853.
de Waele, A.Th.M.H., Reekers, A.B. and Gijsman, H.M. (1976). *Physica* **81B**, 323; and (1976). *Proc. Sixth Int. Cryogenic Engineering Conf.*, Grenoble, France, 11–14 May 1976, (London: IPC Science and Technology Press (UK)), p. 112.
Dokoupil, Z., Kapadnis, D.G., Sreeramamurty, K. and Taconis, K.W. (1959). *Physica* **25**, 1369.
Dow, R.C.M., Guénault, A.M. and Pickett, G.R. (1982). *J. Low Temp. Phys.*, **47**, 477.
Durieux, M. and Rusby, R.L. (1983). *Metrologia*, **19**, 67.
Ebner, C. and Edwards, D.O. (1970). *Phys. Rep.*, **2C**, 77.
Edwards, D.O., Brewer, D.F., Seligmann, P., Skertic, M. and Yaqub, M. (1965). *Phys. Rev. Lett.*, **15**, 773.
Edwards, D.O., Ifft, E.M. and Sarwinski, R.E. (1969). *Phys. Rev.*, **177**, 380.
Ehnholm, G.J., Ekström, J.P., Jacquinot, J.F., Loponen, M.T., Lounasmaa, O.V. and Soini, J.K. (1980). *J. Low Temp. Phys.*, **39**, 417.
Eska, G. and Neumaier, K. (1983). *Cryogenics*, **23**, 84.
Fisher, R.A., Hornung, E.W., Brodale, G.E. and Giauque, W.F. (1973). *J. Chem. Phys.*, **58**, 5584; see also Giauque, W.F., Fisher, R.A., Hornung, E.W. and Brodale, G.E. (1973). *J. Chem. Phys.*, **58**, 2621.
Frossati, G. (1978). Obtaining ultralow temperatures by dilution of ^{3}He into ^{4}He, *Proc. 15th Int. Conf. on Low Temperature Physics*, Grenoble, France, 23–29 August 1978, in *J. de Physique, Colloque C-6, Suppl. no. 8*, p. 1578.
Frossati, G., Godfrin, H., Hébral, B., Schumacher, G. and Thoulouze, D. (1978). *Proc. Int. Symp. on Physics at Ultralow Temperatures*, Hakoné, Japan, 5–9 September 1977, ed. T. Sugawara, S. Nakajima, T. Ohtsuka and T. Usui, (Physical Society of Japan): a, Conventional cycle dilution refrigeration down to 2.0 mK p. 205; b, Miniature plastic Pomeranchuk cells p. 294.
Gachechiladze, I.A., Pavlov, D.V. and Pantsulaya, A.V. (1986). *Cryogenics*, **26**, 242.
Ghozlan, A. and Varoquaux, E.J.A. (1975). *C. R. Hebd. Séan. Acad. Sci.* (Paris), ser. B, **280**, 189.
Greywall, D.S., (1983). *Phys. Rev.*, **B27**, 2747.
Greywall, D.S., (1985). *Phys. Rev.*, **B31**, 2675.
Greywall, D.S. and Busch, P.A. (1980). *Rev. Sci. Instrum.*, **51**, 509.
Greywall, D.S. and Busch, P.A. (1982). *J. Low Temp. Phys.*, **46**, 451.
Grilly, E.R. (1971). *J. Low Temp. Phys.*, **4**, 615.
Gylling, R.G. (1971). *Acta Polytechnica Scandinavica*, Ph**81**.

Hall, H.E., Ford, P.J. and Thompson, K. (1966). *Cryogenics* **6**, 80.

Halperin, W.P., Archie, C., Rasmussen, F.B., Buhrman, R. and Richardson, R.C. (1974). *Phys. Rev. Lett.*, **32**, 927.

Halperin, W.P., Rasmussen, F.B., Archie, C.N. and Richardson, R.C. (1978). *J. Low Temp. Phys.*, **31**, 617.

Harrison, J.P. (1979). *J. Low Temp. Phys.*, **37**, 467.

Hensel, P. (1973). *J. Low Temp. Phys.*, **13**, 371.

Hudson, R.P. (1972). *Principles and application of magnetic cooling* (Amsterdam: North-Holland).

Hudson, R.P., Marshak, H., Soulen Jr, R.J. and Utton, D.B. (1975). *J. Low Temp. Phys.*, **20**, 1.

Huiku, M.T., Jyrkkiö, T.A., Kyyäräinen, J.M., Loponen, M.T., Lounasmaa, O.V. and Oja, A.S. (1986). *J. Low Temp. Phys.*, **62**, 433.

Ishimoto, H., Nishida, N., Furubayashi, T., Shinohara, M., Takano, T., Miura, Y. and Ôno, K. (1984). *J. Low Temp. Phys.*, **55**, 17.

Johnson, R.T., Lounasmaa, O.V., Rosenbaum, R., Symko, O.G. and Wheatley, J.C. (1970). *J. Low Temp. Phys.*, **2**, 403.

Johnson, R.T., Rosenbaum, R., Symko, O.G. and Wheatley, J.C. (1969). *Phys. Rev. Lett.*, **22**, 449.

Johnson, R.T. and Wheatley, J.C. (1970). *J. Low Temp. Phys.*, **2**, 423.

Jurriëns, R.G., Pennings, N.H., Satoh, T., Taconis, K.W. and de Bruyn Ouboter, R. (1978). *Proc. Int. Symp. on Physics at Ultralow Temperatures*, Hakoné, Japan, 5–9 September 1977, ed. T. Sugawara, S. Nakajima, T. Ohtsuka and T. Usui, (Physical Society of Japan) p. 226.

Kobayasi, S., Shinohara, M. and Ôno, K. (1976). *Cryogenics*, **16**, 597.

Koike, Y., Fukase, T., Morita, S., Okamura, M. and Mikoshiba, N. (1985). *Cryogenics*, **25**, 499.

Kopietz, P., Dutta, A. and Archie, C.N. (1986). *Phys. Rev. Lett.*, **57**, 1231.

Korringa, J. (1950). *Physica*, **16**, 601.

Kuerten, J.G.M., Castelijns, C.A.M., de Waele, A.T.A.M. and Gijsman, H.M. (1985). *Cryogenics*, **25**, 419.

Kummer, R.B., Adams, E.D., Kirk, W.P., Greenberg, A.S., Mueller,R.M., Britton, C.V. and Lee, D.M. (1975). *Phys. Rev. Lett.*, **34**, 517.

Laheurte, J.P. and Keyston, J.R.G. (1971). *Cryogenics*, **11**, 485.

Landau, J., Tough, J.T., Brubaker, N.R. and Edwards, D.O. (1970). *Phys. Rev.*, **2A**, 2472.

Lounasmaa, O.V. (1974). *Experimental principles and methods below 1 K* (London: Academic).

Lounasmaa, O.V. (1979). *J. Phys. E.: Sci. Instrum.*, **12**, 668.

Marshak, H. (1983). *J. Res. Nat'l Bur. Stand. (USA)*, **88**, 175.

Marshak, H. (1986). Chapter 16 (entitled 'Nuclear orientation thermometry') *Low Temperature Nuclear Orientation*, ed. N.J. Stone and H. Postma (Amsterdam: North Holland).

McClintock, P.V.E., Meredith, D.J. and Wigmore, J.K. (1984). *Matter at low temperatures* (Glasgow: Blackie).

Mueller, R.M., Buchal, Chr., Folle, H.R., Kubota, M. and Pobell, F. (1980). *Cryogenics*, **20**, 395.

National Physical Laboratory (UK Department of Industry) (1976). *The international practical temperature scale of 1968 – amended edition of 1975* (London: Her Majesty's Stationery Office).

Niinikoski, T.O. (1976). Dilution refrigeration: new concepts, *Proc. Sixth Int. Cryogenic Engineering Conf.*, Grenoble, France, 11–14 May 1976, (London: IPC Science and Technology Press (UK)) p. 102.

Osheroff, D.D., Richardson, R.C. and Lee, D.M. (1972). *Phys. Rev. Lett.*, **28**, 885.

Osheroff, D.D. and Yu, C. (1980). *Phys. Lett.*, **A77**, 458.

Parpia, J.M., Kirk, W.P., Kobiela, P.S. and Olejniczak, Z. (1985). *J. Low Temp. Phys.*, **60**, 57.

Pennings, N.H., de Bruyn Ouboter, R. and Taconis, K.W. (1976). *Physica*, **81B**, 101; and *Physica*, **84B**, 249.

Pobell, F. (1982). *Physica*, **109 & 110B**, 1485.

Reijntjes, P.J., van Rijkswijk, W., Vermeulen, G.A. and Frossati, G. (1986). *Rev. Sci. Instrum.*, **57**, 1413.

Ritchie, D.A. (1985). Doctoral Thesis, University of Sussex. See also Ritchie, D. A., Saunders, J., and Brewer, D. F. (1987). *Phys. Rev. Lett.* **59**, 465.

Robinson, F.N.H. (1987). *J. Phys. E: Sci. Instrum.*, **20**, 399.

Rose-Innes, A.C. (1973). *Low temperature laboratory techniques*, 2nd ed. (London: English Universities Press).

Rusby, R.L. (1985). *J. Low Temp. Phys.*, **58**, 203.

Rusby, R.L. and Durieux, M. (1984). *Cryogenics*, **24**, 363.

Šafrata, R.S., Koláč, M., Matas, J., Odehnal, M. and Švec, K. (1980). *J. Low Temp. Phys.*, **41**, 405.

Sagan, D. (1981). Dilution refrigerators, a section from the privately-circulated *Low temperature techniques – Spring 1981* (probably best to write to Prof. R.C. Richardson, Laboratory of Atomic and Solid State Physics, Cornell University, Clark Hall, Ithaca, NY 14853, USA).

Seppä, H. (1986). *J. Low Temp. Phys.*, **62**, 329.

Satoh, T., Jurriëns, R.G., Taconis, K.W. and de Bruyn Ouboter, R. (1974). *Physica*, **77**, 523.

Satoh, N., Satoh, T., Ohtsuka, T., Fukuzawa, N. and Satoh, N. (1987). *J. Low Temp. Phys.*, **67**, 195.

Soulen Jr, R.J. and Dove, R.B. (1979). *National Bureau of Standards (US Department of Commerce) Special Publication 260–62*, entitled 'Standard reference materials: SRM 768: temperature reference standard for use below 0.5 K'.

Soulen Jr, R.J. and Giffard, R.P. (1978). *Appl. Phys. Lett.*, **32**, 770.

Soulen Jr, R.J. and Marshak, H. (1980). *Cryogenics*, **20**, 408.

Sullivan, N.S. and Edwards, C.M. (1986). *Cryogenics*, **26**, 211.

Taconis, K.W., Pennings, N.H., Das, P. and de Bruyn Ouboter, R. (1971). *Physica*, **56**, 168.

Vermeulen, G.A. and Frossati, G. (1987). *Cryogenics*, **27**, 139.

Vermeulen, G.A., Wiegers, S.A.J., Kranenburg, C.C., Jochemsen, R. and Frossati, G. (1987), *Can. J. Phys.*, **65**, 1560; see also G.A. Vermeulen's doctoral thesis, 1986 (Leiden: Holland).

Vvedenskii, V.L. and Peshkov, V.P. (1972). *Zh. Eksp. Teor. Fiz.*, **63**, 1363;

English translation in *Sov. Phys. J.E.T.P.*, **36**, 721 (1973), which has pages 722 and 723 confusingly interchanged.

Webb, R.A., Giffard, R.P. and Wheatley, J.C. (1973). *J. Low Temp. Phys.*, **13**, 383.

Wei-chan Hsu and Pines, D.J. (1985). *J. Stat. Phys.*, **38**, 283.

Wheatley, J.C., Rapp, R.E. and Johnson, R.T. (1971). *J. Low Temp. Phys.*, **4**, 1.

Wheatley, J.C., Vilches, O.E. and Abel, W.R. (1968). *Physics*, **4**, 1.

White, G.K. (1979). *Experimental techniques in low temperature physics*, 3rd edn (Oxford: Clarendon Press). Note that the 1987 paperback reprint has corrections and extra references.

Wilks, J. and Betts, D.S. (1987). *An introduction to liquid helium*, 2nd edn (Oxford: Clarendon Press).

Index

adiabatic demagnetisation
  general coverage 7, 60–80
  model calculations 65–9
  geometry of the refrigerant 70–3
  apparatuses 73–80
Allen–Bradley resistors as thermometers, *see* carbon

capacitance for thermometry, *see* glass
carbon (C), electrical resistance of 87–9
Carnot cycle 11
cerous lanthanum magnesium nitrate (CLMN)
  magnetic susceptibility of 81, 89–92
  as refrigerant 60
  deuterised 60
cerous magnesium nitrate (CMN)
  entropy of 8
  magnetic susceptibility of 89–92
  as refrigerant 1, 60
chromic potassium alum 60
circulation rate of a dilution refrigerator 40–1
CLMN or CMLN, *see* cerous lanthanum magnesium nitrate
CMN, *see* cerous magnesium nitrate
cobalt-60 (^{60}Co) 81
copper (Cu)
  entropy of 62–4
  as refrigerant 1
  as thermometer using NMR 91–4
cryostats ('pots'), ^{3}He and ^{4}He 25, 26

dilution refrigeration
  conventional method 24–30
  alternative methods 43

EPT-76, *see* International Practical Temperature Scales
evaporation cooling 24

Fermi degeneracy of solute ^{3}He 17, 37
film burning or inhibiting still, *see* still
film flow suppression 32

free expansion 1

gamma-ray anisotopy thermometry 81, 94
germanium (Ge), electrical resistance of 87
glass (SiO$_2$), electrical capacitance for thermometry 89–90

heat exchangers (of a dilution refrigerator) 34–43
helium-3 (^{3}He)
  melting, properties of 48
  as refrigerant 1
  superfluid transitions 13
  melting pressure 14, 81
  phase diagram 15, 48, 50
  cryostats ('pots') 25, 26
  properties of 29
  entropies of liquid and solid on melting curve 47
helium-4 (^{4}He)
  as refrigerant 1
  melting pressure 14
  superfluid transition 14
  phase diagram 15
  cryostats ('pots') 25
^{3}He-^{4}He mixtures
  as refrigerant 1
  dilute 17
  dilute, saturated, properties of 29
  specific heat 19, 29
  osmotic pressure 18–20, 29, 31
  vapour pressure 21
hyperfine-enchanced Van Vleck paramagnets 69

International Practical Temperature Scales 12, 84
IPTS-68: *see* International Practical Temperature Scales
inversion curve 5
inversion temperature, maximum value of 7
isenthalpic expansion 5

isentropic expansion 2

Kapitza resistance 34, 39
Kelvin scale 10, 81
Korringa mechanism 66

Leiden refrigerator 44–5

Matsushita resistors as thermometers, *see* carbon
mixing chamber (of a dilution refrigerator) 27
melting pressure
    ^{3}He 14, 47, 81, 85
    ^{4}He 14

NBS superconducting fixed point device 83
noise thermometry 95
nuclear magnetic resonance (NMR) 91–4
nuclear orientation thermometry 81
nuclear paramagnet (ideal) 64

osmotic pressure of ^{3}He dissolved in ^{4}He 18–20, 31

phase diagrams: ^{3}He, ^{4}He, and ^{3}He/^{4}He mixtures 14, 15, 48
phonons 18
platinum NMR thermometry 81, 91–4
Pomeranchuk refrigerator 46–59
pots, *see* cryostats
praesodymium nickel five (PrNi$_5$)
    entropy of 62
    as refrigerant 1

Robinson oscillator 90–1
rotons 18

specific heat
    ^{3}He/^{4}He mixtures 19, 29
    ^{3}He 29
Speer resistors as thermometers, *see* carbon
spin–lattice relaxation 66
still (of a dilution refrigerator) 31–3
superconducting transitions 13
superfluid transitions
    ^{3}He 13
    ^{4}He 14

thermal conductivity, *see* transport properties
thermometry
    general coverage 10, 81–95
    temperature ranges of various types 82
third law of thermodynamics 17
transport properties: ^{3}He/^{4}He mixtures 21–3
two-fluid model for ^{3}He/^{4}He mixtures 18

Van der Waals fluid
    internal energy of 2
    entropy of 3
    inversion curve of 5
    enthalpy of 6
Van Vleck paramagnets, *see* hyperfine-enhanced Van Vleck paramagnets
vapour pressure
    ^{3}He 84
    ^{3}He/^{4}He mixtures 21
viscosity, *see* transport properties
viscous heating in heat exchangers 42

Wiedemann–Franz law 66